Wildlife testimonials

"Volunteering under Ann was a life enriching experience. She is a true ambassador of the health and well being of animals"

Karen Malone

The following is a poem Karen wrote and gave to me in a frame when I left the center.

Wildlife

They are less frightened by her touch and are
Trustful of her actions

To all the animals she shows her soul
She soothes and heals, comforts and cares

When unable to help, she lovingly points the way
to eternity.

How fortunate so many have known her

Karen Malone

"Ann Southcombe has an incredible ability to empathize and communicate with nonhuman animals. A passionate advocate of animal protection, she has enhanced the lives of many precious beings over the past forty years."

Katherine Oxendine, Founder and President,
The Toby Fund of Wolf Creek Oregon, Inc.

"Beyond a good storyteller, Ann Southcombe has the rare and remarkable gift of letting the reader hear animals' own voices. These rich accounts provide a beacon for humanity as we make our road to inters-species compassion and peace."

G.A. Bradshaw Ph.D., Ph.D.
Founder and Director of The Kerulos Center
and author of Elephants on the Edge

A Wild Life

Healing animals with an open mind and loving heart

Ann Southcombe

Book Publishers Network
P.O. Box 2256
Bothell • WA • 98041
Ph • 425-483-3040
www.bookpublishersnetwork.com

10 9 8 7 6 5 4 3 2 1

Printed in the United States of America

LCCN 2010916712
ISBN10 1-935359-70-3
ISBN13 978-1-935359-70-8

Editor: Julie Scandora
Cover Designer: Laura Zugzda
Typographer: Stephanie Martindale

Disclaimer

The views expressed in this book are the views of the Author and do not necessarily represent the views of the publisher or of any wildlife rehabilitation education center. Caring for wild animals and/or keeping wild animals in captivity may be against various federal, state or local wildlife laws. Such actions may be legal but may require special licenses or permits. For example, the Author has an Oregon animal rehabilitation holding permit. Before you decide to care for wild animals, you should check to make sure you are in compliance with applicable laws and regulations.

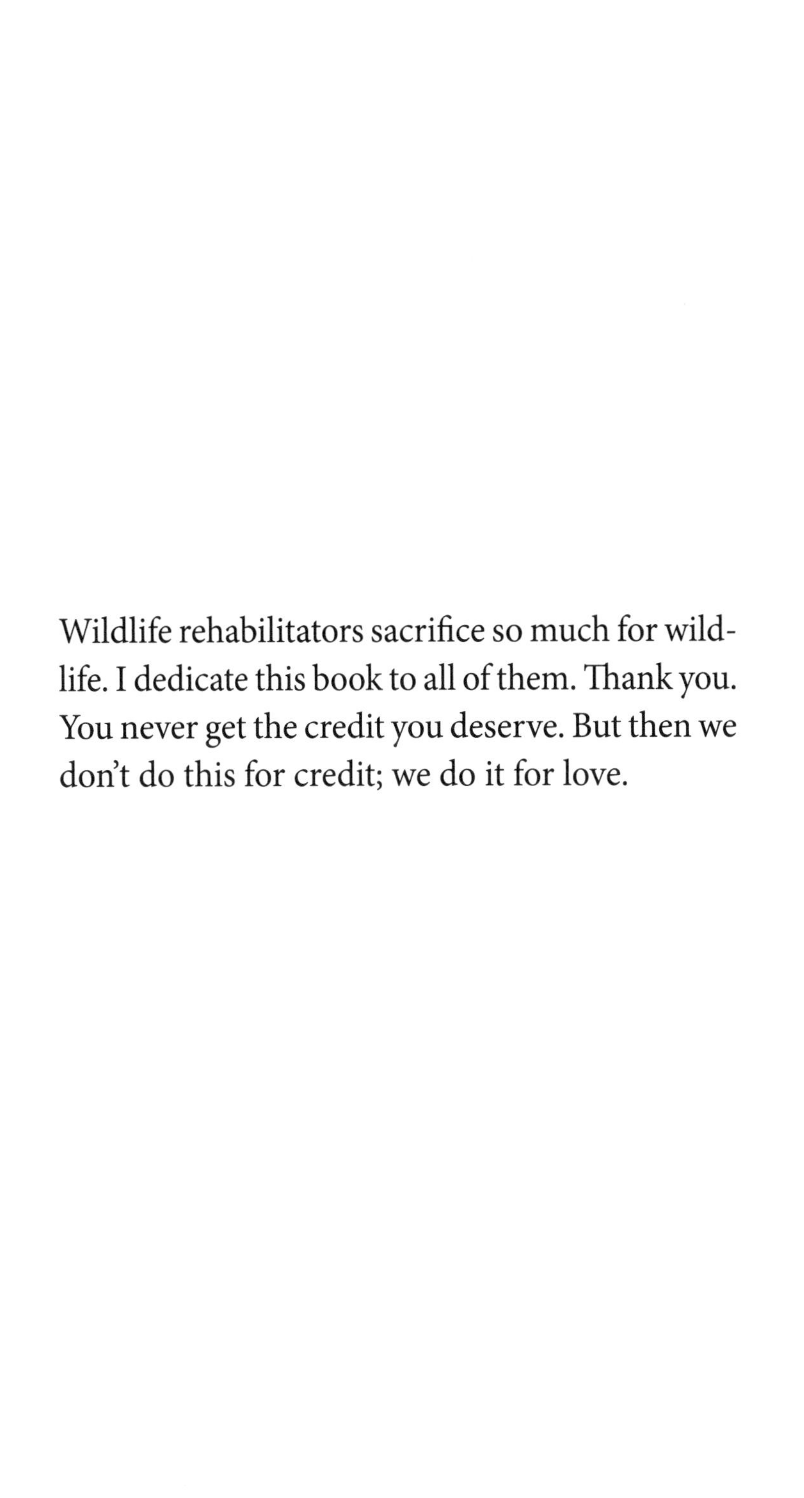

Wildlife rehabilitators sacrifice so much for wildlife. I dedicate this book to all of them. Thank you. You never get the credit you deserve. But then we don't do this for credit; we do it for love.

Contents

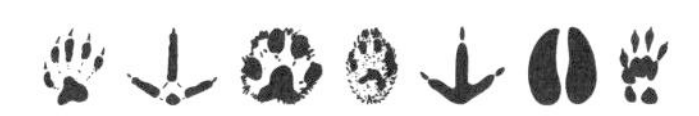

Introduction

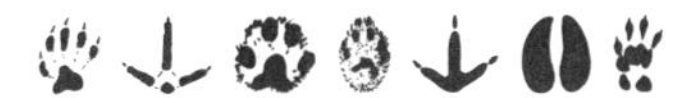

This book contains excerpts from my larger book, *From Gorillas to Squirrels: One Woman's Life Journey in Kinship with Animals*. My life's journey has been made in many smaller journeys or stages—each a place I worked, relating to different animals and learning from them.

My seventh journey, the one told here, was the most intense learning experience of all—not just physically and emotionally but spiritually. Native wildlife are the non-human animals we share our home territories with, and these experiences brought me closer both to truths about what our human nature consists of and to our connectedness to the wild.

Along with these animal experiences, I also learned that being a wildlife rehabilitator is one of the hardest yet most rewarding jobs on the planet. Many would say that saving injured and orphaned wildlife is not a productive occupation and doesn't make a difference. I say it makes a *big* difference. For me, it is one of the things that keeps me from disliking the human species. It takes all the best qualities of being human to be a rehabber—compassion, patience, extraordinary stamina, and enough love to risk and endure extreme heartache.

I have met incredible rehabbers on “The Squirrel Board,” an Internet message board (www.thesquirrelboard.com). Many of these people are saving wildlife out of their homes with no monetary compensation, worrying about the injured and sick, feeding babies every two hours, and spending many sleepless nights. Most of the animals coming to them have been injured or abandoned and are at death’s door from the start. It is a given in this job that rivers of tears will flow. But thankfully, when an orphaned or injured animal is released back to nature and its natural life, joy takes away the tears.

Acknowledgments

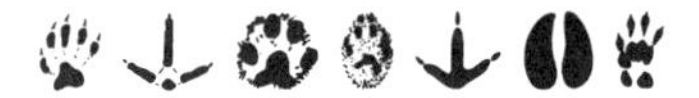

I am grateful to all the dedicated volunteers who helped me at the wildlife center. Many became my family and are still in my life.

I also wish to thank the Southern Oregon Writers Group along with Chris, my friend and editor. This book would never have happened without their help.

Photo credits: Ann Southcombe, Denise Cutrell, and Sheila Kaufman (thank-you)

Chapter 1

Boo Boo and The Bear

All my animal careers have included one-on-one animal relationships. In zoos and sanctuaries, the animals I cared for were mostly permanent residents, and attachment to some extent could be mutually beneficial. But in wildlife rehab, I help heal or raise them, only to send them back to nature. In my rehab work, I've had to learn to let go.

In most cases, the animals don't stay with us long enough for strong emotional bonds to form on either side. The orphaned babies start out as cute individuals, but after being weaned, they're placed in pre-release cages with others of their own species. At this stage, they must learn natural survival skills before being returned to the wild. It is essential that they do not become too friendly with humans, and usually we can work within that limitation. But once in a while, an animal comes in that I immediately connect with even though it is wild and fearful of me. This isn't something I make a decision about; it just happens. For example, in my first year at the center, some orphaned baby bears crawled right into my heart. By the time they were released a year later, my feelings were definitely mixed: I was happy they were free yet very sad they were leaving my life.

One summer a few years later, a similar thing happened when a two-month-old bobcat baby was brought to us. Its mother had been seen too close to a house, and the owner shot and killed her. The next day the same family found her baby wandering around. Luckily for this cub, they felt pity and brought him to us. He was so tiny and scared that immediately I felt empathy for him for losing his mom. I didn't just know but *felt* that he would no longer have her to cuddle with or to groom him gently with her warm tongue. And growing up, he wouldn't have siblings or a mother to play with. This bobcat's natural loving home was gone. I knew I didn't want him to live in isolation at our facility until he was old enough to release. Yet I didn't want him to become too human-friendly either. I tried to compromise and hoped for the best.

The baby bobcat was given an area in our quarantine building a few yards from the clinic. Inside, there were three smaller rooms, with a fenced-in outdoor area off the back of each. These inside rooms were approximately four feet wide and five feet long with seven-foot tall walls, all lined with white plastic sheeting. There were no windows, but a small door led to a larger outside cage with chain-link fencing. Indoors, he had blankets on the floor, a litter box, and a sleeping box for a den. Outside was the "wild." Outdoors, the floor was covered with tree bark mulch and pine needles. Looking forward to helping him develop his climbing skills, we added a log and several branches in tree-like positions.

Every night after work, I sat in the cub's enclosure and talked softly to reassure him I was safe. While he was getting used to me, I stayed only fifteen or twenty minutes, which seemed to be the amount of time he could tolerate without becoming stressed. He darted in and out of the inside cage where I sat but stayed mostly in his outside cage. After five days of this, he decided to stay with me a while in the same space, and then I started bringing a small stuffed teddy bear for him to play with.

Boo Boo

This bear was about six inches tall, brown, and furry. My intention was to use it as a combination toy and prey. Play is a very important part of a baby animal's life. Not only is it fun—it also leads the animals to develop many survival skills. Hunting and killing skills develop during play when predators are young, and I was hoping the little bear could take the place of a sibling to play with. I moved it around slowly to get the bobcat's attention—always wearing leather gloves to protect myself. If he missed the bear and got part of me, instead, I didn't want to inhibit any of his natural behavior by scolding him for biting. As I moved the Boo Boo bear, as I called it, the bobcat would stalk and pounce on it. After a few nights of such play, I started calling the bobcat himself "Boo Boo," too.

As time passed, I noticed the bobcat's skills improving—and my emotional attachment getting stronger. I so wanted him to trust me enough to let me cuddle and pet him, but I knew that would not be in his best interest. I think I must have found the balance I sought, because he trusted me enough to come in and play near me, but he never wanted actually to be with me. Occasionally, he would jump up and grab the bear, then land on my leg or lap. I believe this form of touch made him feel a protective presence that increased his sense of security without turning into an attachment that would hinder the independence he would need later on.

One night, I came in to see Boo Boo after I'd been working with an injured bird and still had that smell on my shirt. He sensed it right away and hopped up on my shoulder as I sat in the corner. He began rubbing his head near my neck over and over, as if the bird scent were catnip. I was a bit tense—this animal was not tame and occasionally nipped at me. I had some fear he might suddenly realize how close he was to me, chomp on my ear, and run! Luckily, he was so enraptured with his bird experience he didn't seem to know where he was. Before he came out of his trance, I slowly picked up the bear and tossed it to the other end of the room. That broke the spell, and he took off after the "prey." Whew!

Boo boo enraptured, me concerned

BooBoo the bobcat with his toy bear

As the months passed, Boo Boo grew bigger and needed more space. On one side of his inside room was a siding door that opened to an adjoining cage, and we made that area into another "wild room" with branches and mulch. He graduated from eating his meat from a bowl to chasing and catching it when we tossed it for him. Then the time came for him to disassociate humans from food, and we started hiding it throughout his "forest" for him to find later. Meanwhile, still hoping to improve his hunting skills, I continued his prey training with the bear. Whenever I approached his door, he made a fierce growl. This made me feel good because it told me he still had a healthy fear of me. Some wildlife rehabilitators prefer to put live rabbits, chickens etc. in with young Predators like Boo Boo. I couldn't do this nor do I think it is necessary. My domestic Cats are very good hunters at a young age, I am sure a much wilder one is equal if not better.

When I saw that the growing bobcat needed to start using more of his running muscles, I began letting him out into the thirty-foot-long hallway adjoining his cage so I could toss the bear for him to chase. As he went through the opening into the hallway, he growled ferociously—until he started playing with the bear. He was young, after all, and this was still play to him. Then the game began. With him at one end of the long hallway and me at the other, I held the "prey" in my hand and moved it along while he slowly crawled toward it on his belly, coming closer and closer. Just as he was about to pounce on the bear, I'd toss it over his head. He'd leap high in the air and catch it with his claws, then his teeth. Then off he'd run with his prize to the other end of the hallway. Once there, the bear was "dead," and he'd lose interest in it. By now, I had several prey animals to keep him on the run. One of his favorite things to chase was a golf ball. I would roll it, and once he caught it, he would bat it around so he could continue to chase it over and over. His play was just like that of my cats at home, only he was about three times their size.

Boo Boo a few weeks before his release

Some nights, seeing the bobcat's cute young face lighting up as he caught his phony prey, I got tears in my eyes because I knew that someday this wouldn't be in fun. My little Boo Boo would be out there competing for food with coyotes, cougars, and even other bobcats. Human predators were also a force to fear. They could hunt and kill him for just being himself—a bobcat.

The time was quickly approaching for the young animal to be released. He was nearly seven months old, it was April, and Spring was beginning. The weather was good, and there would soon be plenty of baby animal out there for him to find for food, many of them as inexperienced in evasion as he was in pursuit. It was the perfect time to set him free.

At our center, when a release is imminent, a date is usually scheduled so that TV stations can cover the story and a select few can be invited to see it happen. It is truly thrilling to see an animal go free that otherwise might not have had a chance to live, and these are popular events. We tried and tried to find a time when all the interested parties could attend, but no one could come up with a date. And Boo Boo was more than ready; he was getting too big for his cages—and bored, as well. Finally, I decided the time had come for action and asked Kat, a volunteer who'd been helping with Boo Boo, to help. We crated him up, and the two of us drove off with him to the wilderness, searching for a good place with water nearby and as far as possible from humans. Of course, it's impossible to guarantee a perfect location; you can't be sure there will be food, or that you aren't leaving the animal in another's territory, where it will be chased away. You just do the best you can.

We pulled off the main road onto a forest road and drove a few more miles. Coming to an open area surrounded by brush and trees with a creek nearby, we stopped and unloaded Boo Boo's crate onto a grassy patch. When we opened the crate, he remained inside at first. After all, he probably didn't remember his birthplace and only knew his small, human-made home. But after five minutes or so, one paw came out and touched the grass. Then slowly the rest of his body emerged, and he began looking around. There were several young saplings nearby, and he started rubbing himself against them. We threw a dead rabbit and some other pieces of meat around so that he would have food for a while in case he didn't find his own right away, but he paid no attention even though he hadn't eaten that day. Clearly this new world was more important to him just then.

Fifteen minutes or so passed as the young animal explored the nearby area, gradually going further off to the thick brush to explore. And then he scampered out of sight. I felt a twinge of sadness then, knowing how hard life is in the wild and that he was now entering that world, but I wanted to feel happy that he was free to be who he was. I called a soft "Bye, Boo Boo."

To my surprise, he came back. But that was just for one last look, and then he took off forever. Had he come back to say goodbye to me, too? Whatever caused him to look back, it did make me feel happier to see him that one last time.

I felt … I truly knew … he would be fine.

At my home, I later placed the Boo Boo bear in a prominent place in my house surrounded by pictures of Boo Boo playing with it. Like the Velveteen Rabbit, a stuffed toy that helped a little boy in growing up, this bear had had a special job—it had helped a bobcat in learning to survive in the wild.

CHAPTER 2

SHASTA: MY MAJESTIC ONE

A call came in to our rehabilitation center that a bald eagle was lying at the side of the road just a few miles away. I stayed at the clinic, working, while the director, several staff members, and my supervisor, Chris, jumped into our van and drove off to the rescue.

Many times when we got calls about injured eagles, they turned out to be hawks or something smaller, but this time it really was a mature bald eagle. The director lifted the weak bird off the road and handed it to Chris, who gently held it as they drove back to the clinic. There, Chris was able to check out the bird's injuries. At first it looked as if she had been shot in the face, but with closer observation it seemed much more likely that a vehicle had hit her (yes, Chris determined that the eagle was female). On the road where she'd been found, we knew, there was a lot of construction truck traffic. She had probably been at the road's edge eating carrion when a truck came along and clipped the top of her beak. Looked at this way, it seemed amazing that more of her face had not been injured. Her lower beak, eyes, and nostrils were all intact—but of course, without the top beak, she would not be able to eat on her own, and without help, she would starve.

We also observed that many of her chest feathers were missing and knew that meant she was probably nesting because she would have used those feathers in her nest. Several volunteers rushed back to the accident site to search for a nest in the area, in case we could save abandoned babies or eggs, but they didn't find one.

By the time we got to the bird, she had probably been on the side of the road for quite a while, for she was very cold and weak. After cleaning the wound to her beak, Chris put her in a large animal carrier with a heating pad. Once her body temperature was brought back to a safe level, Chris began to give the eagle warm fluids through a feeding tube. This was a small syringe with a thin metal tube attached to the end, which slid down her throat directly to her stomach—she was too weak to swallow. Chris fed her liquids every two hours through the night, and the next day we began adding strained baby food meat to the fluids.

For two days, we followed this routine: after carefully taking the top off the carrier, we wrapped a sheet or towel over the bird's body (this helped her feel secure and kept her calm for these feedings) and fed her twelve ounces of the mixture of liquids and baby food, repeating the process three times a day. As the patient gained in strength, our vet felt it would be better for her if we modified the feeding tube procedure. He suggested that a plastic tube could be inserted down her throat into her stomach, routed from her mouth over her shoulder, sutured to her back, and left in place. That way we wouldn't have to put the tube down her throat every time for feedings; we could just put a towel over her head to relax her, and then push her food down the opening secured to her back. Once we set up the feeding the way the vet had suggested, the eagle barely seemed to know when we were feeding her, and it was much less stressful for all of us.

Each day, the eagle got a little stronger, and after two weeks of tube-feeding, she was ready for solid food, and we could remove

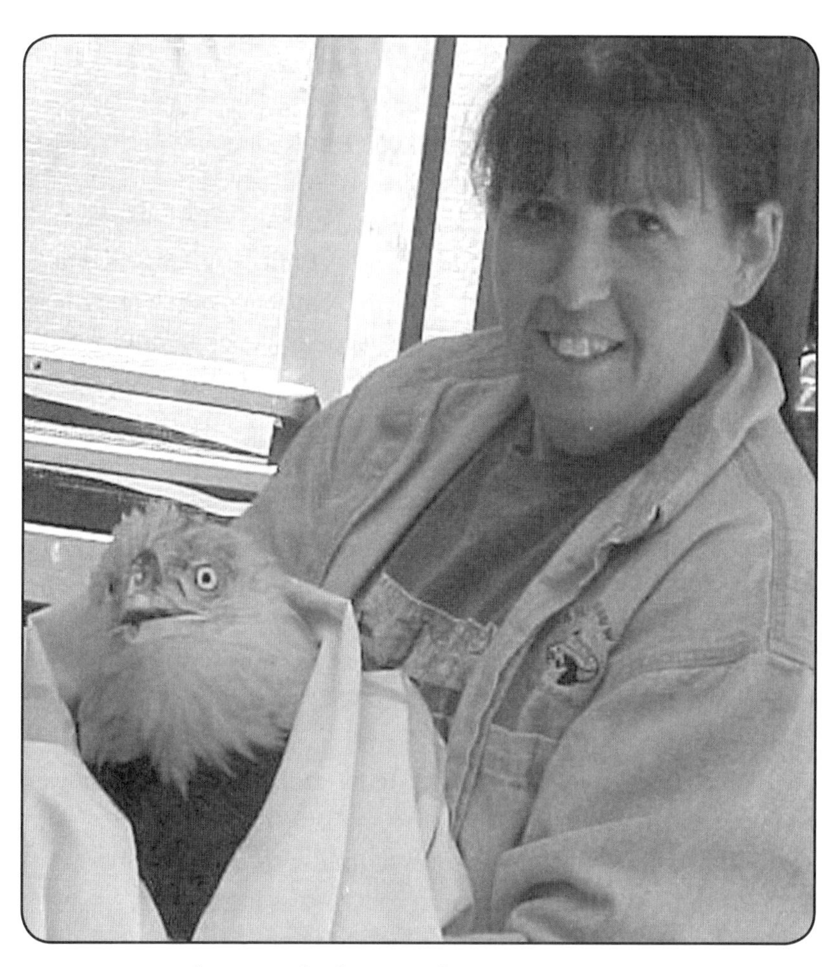

Shasta and Chris on the way to surgery

the tube apparatus. Now she had a new procedure to get used to. After calming her with the sheet over her body, we had to open her mouth and stick a piece of meat down her throat with a large pair of tweezers. She seemed to accept this routine with the same resignation that she had accepted the feeding tube, but it was awkward for us.

By this time, even though her temperament was incredibly calm for a wild bird, she had become too strong and active to be kept in the animal carrier any longer, so we moved her to a larger enclosure in our quarantine building. Now she lived in a roomy cage, four feet on each side and seven feet tall. Though still small, it was large enough to allow an eagle to stand on a perch and move around. She was still healing, so we needed to keep her as inactive as possible, yet she also needed to be allowed some range of movement as she recovered.

At her feedings, we continued the practice of putting the towel over the eagle's body to calm her because it still seemed to make her feel secure. Probably the regular routine was comforting to her, as well. Now, as soon as we uncovered her face she would open her mouth to be fed, like a baby bird. And finally we were able to use our fingers instead of the large tweezers, for she could swallow on her own when we gently placed the piece of meat in the front of her mouth.

Chris and I felt this bird needed a special name for she was a very majestic, yet endearing personality. We brainstormed and wrote a list of about ten names. We finally settled on "Shasta." It seemed to fit her because Mount Shasta was a mystical, majestic mountain nearby—a spiritual place for many ancient cultures, including Native Americans today.

Both our vet and our director were sure something could be done about Shasta's missing beak. The vet contacted a local dentist who had worked with various materials making false teeth implants for people. He was excited to be a part of the project, and after they all brainstormed possible solutions, he felt confident he could use this same technique to attach Shasta's prosthetic beak. Once they had a few theories to work with, they were ready to have a model built of Shasta's face, and needed to gather more information about her remaining bone structure. As the word spread in the medical community about this unusual challenge, many others expressed a desire to help as well. The local hospital let us do an MRI (after hours), and this gave us a state-of-the-art picture of every detail of the skeletal structure of the eagle's head. She was an amazingly good patient through the MRI. With the vet, Chris and I helped strap Shasta into the part of the machine where she had to lie down, and as she rode through the screening area of the machine she lay very still, even though she breathed heavily. The vet, always conscious of the animal's welfare, was prepared to stop the procedure if Shasta became too stressed, but she completed the test without a problem. This was just the beginning of what would be months of medical activity: investigations, the invention of a novel prosthesis, and four surgeries and recuperations.

As the weeks passed, I started learning more about this bird and her behavior. I began to feel that Shasta and I were forming a bond, and I thought she might soon trust me enough to hand-feed her without having to put the towel over her. I was naïve about raptors and didn't know the towel was not only to calm her but also to protect me in case she became frightened. Bald eagles have the ability to use their talons to grab and carry prey in their feet with one thousand pounds of pressure—quite enough to crush my arms! But as I didn't have this knowledge, I had no preconceived fears either. I gained Shasta's trust a little bit each day by slowly reaching toward her mouth with her food. At first I stayed as far from her as I could be yet still reach her beak. She trusted me, but only at arm's length. If she moved away, I would then towel her, but it wasn't long before she stayed in place and took the food quietly.

After Shasta finished her meal, I would stay with her for a while just to give her company and to give us both time to get to know each other better. I can feel what most animals feel after I spend

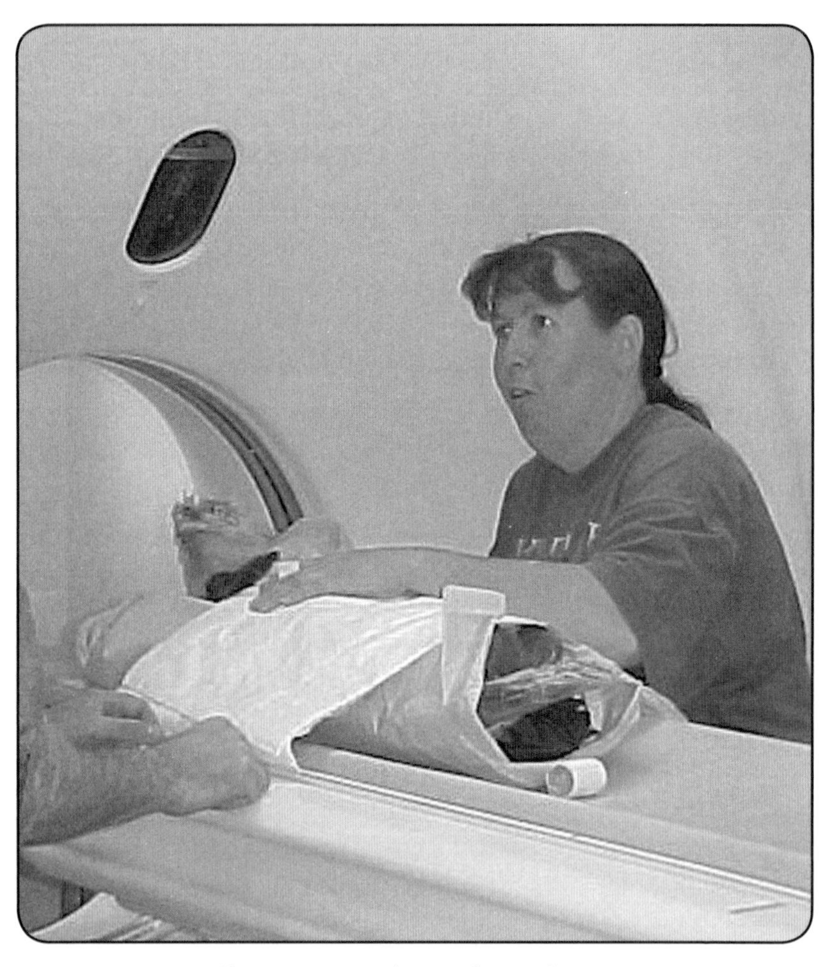

Chris putting Shasta through MRI

quality time with them, but Shasta was still a mystery. For me, just being with any raptor was a foreign experience, and the bald eagle was at the far end of that scale. As I read about the different types of eagles, I learned that golden eagles seem to have a keen awareness of their surroundings, and I thought I could understand this: they need precise environment-reading skills to survive because they arc hunters. The bald eagle, though, has an easier time of it and survives more by catching fish and finding dead animals, which don't require such a high level of attentiveness. I had already noticed a difference in behaviors between Cheyenne, our golden eagle, and Shasta: Cheyenne seemed more intense in her day-to-day behaviors; I could see her "thinking," but I didn't see that going on with Shasta, I didn't know what really was going on with her.

Each day, I tried to test our bond by sitting a little closer or by trying to touch her. The experts had told me that certain animals don't see touch as a positive thing. For instance, for a bird, being touched or petted can trigger the fear of being caught by a predator—especially if among their own species they don't groom and touch each other in stroking ways. And I was told that eagles did not stroke each other. Yet tortoises, another species that does not touch or stroke or mutually groom, does like being petted and stroked. (I had discovered this in one of those surprising incidents when, through a simple response, an animal will completely overturn an assumption we hold about it. Working with giant Aldabra tortoises at a zoo, I had petted one's head as she was chopping away at her food on the ground. Realizing that was a silly thing to do, I stopped petting her—and to my surprise, she raised her head to reach it to my hand again. Then I tried scratching her neck, and again she stretched her head to my hand, to get more stroking.) I didn't know how this eagle would feel about being touched, but Shasta seemed to be okay with short, preening pets on her chest. Then, one day I must have overstepped her boundary, for she reached up with her foot and gently pushed me away. Later I learned how amazing this was, this gentle correction. If I'd known then how powerful an eagle's feet were, I might have panicked when her talons touched my hand, but fortunately, I remained calm and thus didn't frighten Shasta either.

Meanwhile, the medical procedures continued. After Shasta's first surgery, the vet put a protective cone around her neck to prevent her from knocking or rubbing her head on anything. He implanted screws in her head and nose area, hoping that they would be accepted by her bones and secure themselves, because then they could anchor a prosthetic beak in the future. These had to be untouched for six weeks to allow the bone to grow around them. Back at the clinic, Shasta looked very sad and depressed as she stood on her perch, head hanging down to the ground, surrounded by the opaque cone. She wasn't yet used to the added weight of the cone and couldn't hold her head up. We had to go back to covering her with a towel during her feedings, and I took this opportunity to give her extra TLC because, while the towel was over her, she felt secure enough for me to stroke her neck. I gave her little massages, and this relaxed her so much that she usually fell asleep resting against my arm. We all felt much better when, after a week with the cone, she became used to the weight and could hold her head up again—and no longer looked so sad. She was a real trooper and seemed to be able to adjust to just about anything that fate threw at her.

When the time came to remove the cone, we all held our breath, hoping all three screws were securely attached to her bones. The one nearest the tip of her beak seemed to be the strongest, but sadly, the others appeared to have failed. We were so disappointed at this; it meant yet another surgery and recoup time for Shasta. As time and surgeries went by, Shasta became an "on-hold" animal in my life. Our relationship couldn't progress beyond feeding and talking, yet I had great plans eventually to be with her in a more normal eagle-to-caregiver/friend bond. During this time, I investigated new enclosures for her because I wanted her to live out her life in the best circumstances possible in captivity, and I worked at

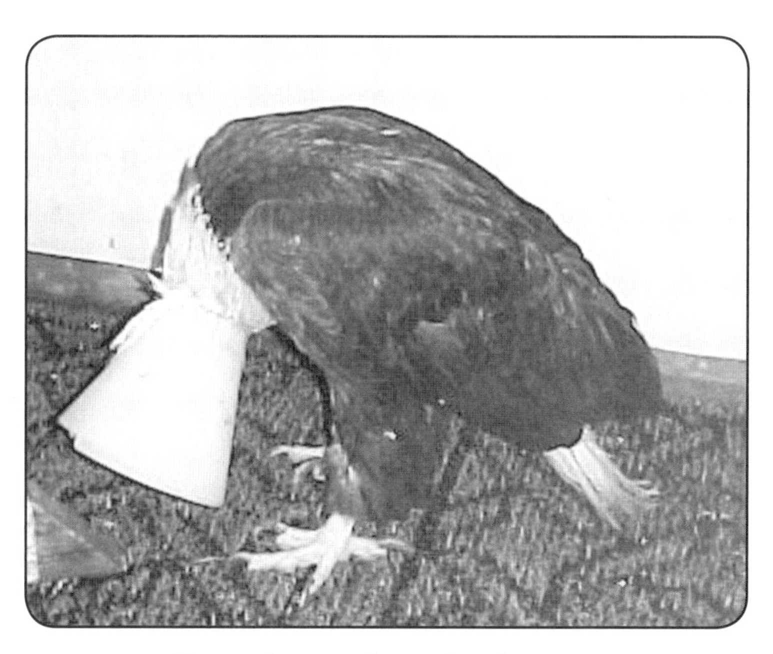

Shasta depressed wearing the cone

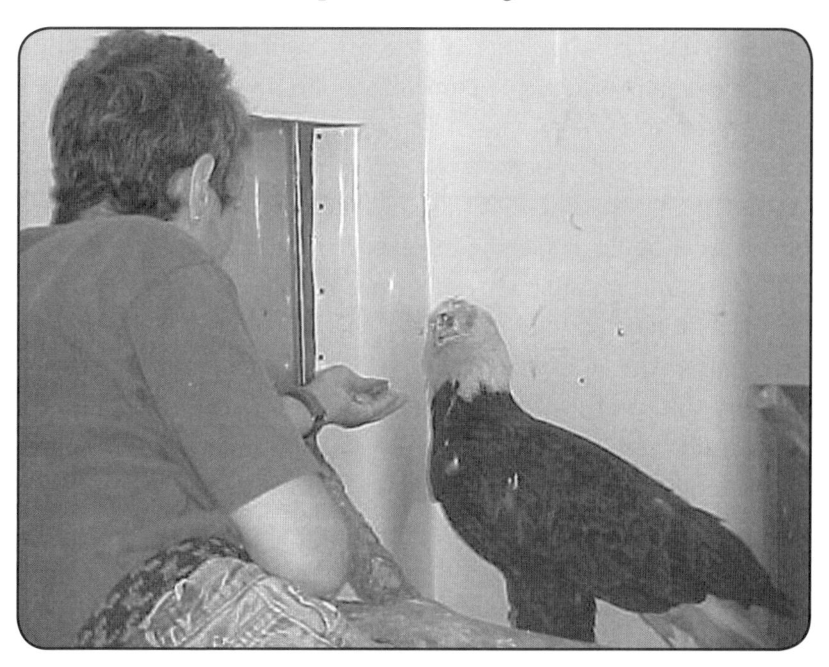

Me hand feeding Shasta

getting materials donated so that we could build something great for her by the time her new beak was in place. Until then, her small enclosure was best for her feeding and recovering after surgery.

We also were thinking of putting jesses on Shasta's feet. These are leather ankle bracelets with straps attached, used since ancient times with trained raptors. Jesses would allow us to tether her on a perch outdoors, where she could get the benefit of fresh air and sunshine. Jesses would also give us the opportunity to train her to sit on a glove on a human's arm so that she could go for walks and, someday, be part of educational programs teaching others about the magnificent bald eagle.

But we had to put this plan on hold for a time, because she was quickly scheduled for a second surgery. Finally, after that recovery, we were able to begin her training with jesses and gloves. Most birds take quite a while to adjust to sitting on a glove; it scares them, and they jump off repeatedly. But when Chris set Shasta on the glove she only jumped off twice before settling down and sitting comfortably. Chris, feeling happy at this, walked her several yards to a bow perch behind the clinic. This perch was about a foot off the ground and got its name from its shape (it was curved, like a rainbow). Shasta hopped off the glove onto the perch, and Chris secured her with a four-foot leash. She tried to fly a couple of times, but soon enough she seemed to understand that she couldn't go anywhere, and settled back on her perch.

Shasta seemed to enjoy the sunshine that day; it was the first she'd felt or seen it in many months. She stood tall and fluffed her feathers, and then, with an alert stare, she surveyed the area. Of course, although she could fly, we could never release her because, even though she could catch prey, she could not tear the flesh with her missing beak. Eventually she would die of starvation. A few days after this first day out, our director took Shasta for a walk on his gloved arm, accompanied by a reporter and cameraman from the local TV channel, who would report the event on the news.

Once more, Shasta took it all with majestic calm, as if she were the subject of media attention every day.

After many months of interaction, Shasta and I had a kind of routine. When I stopped by to check in on her, she would talk to me with her eagle greeting, a sort of chirpy, high-pitched "caw caw caw." She showed a healthy eagerness about her food, attempting to grab it on her own even though, without a top beak, she always needed our help with that.

But as Shasta approached her second year with us, I began to see signs that she might be losing heart. I noticed that she didn't seem quite as interested in eating as she usually was; too often, I would have to open her mouth and stick the food down—and then she would sometimes spit it out. This was unusual. I tried all her favorite foods—raw chicken, venison, and salmon—and still she was increasingly reluctant to eat. Now, after each feeding, I would spend extra time with her, offering pep talks about not giving up.

At this point in our relationship, when we had these talks, Shasta was allowing me to come within inches of her face. Occasionally she looked at me, but most of the time she just looked around the room. She didn't ever move away, so I knew she was not averse to my being there with her. Her body language became more relaxed so I believed she probably even liked this kind of company.

In our big flight cage about a hundred yards away from Shasta's small one, there was an injured bald eagle named Piqua who often vocalized. When I imitated these eagle sounds, Shasta would look at me, and I'd say, "That's Piqua." There were also some ravens nearby, and when they squawked and Shasta reacted to that, I would tell her about them. Sometimes she would respond to my vocalizations with a noise of her own. So I guess you could say we had some conversations about the outside world. Who knows what was going on in her brain? I do know it made me feel better to know that Shasta was receiving some attention and kindness.

After about a month of these pep talks and conversations on topics of interest to eagles, Shasta finally began to show some enthusiasm about her food again. I felt she must have been coming out of her depression and was determined not to give up! She even began actually staring at me in the eyes for long periods of time. This was something she had never done before, and made me wonder even more, "What are you thinking, Shasta?"

Then came the time for Shasta's fourth surgery. Her doctors knew that the device they had originally planned for her would not work, so they had to go back to the drawing-board. During this operation, they were going to clean out her second nasal passage to help her breath better. They would also make a new mold of her head—and, they hoped, with this new information, they could come up with a plan for her beak that would work.

On the day of this surgery, Chris and I got Shasta ready for another trip to the vet. By now, we knew it was best for her and the doctors if we wrapped her feet comfortably with "vet wrap," a wide, expandable bandage. This protected anyone handling her from being jabbed by her talons, and she didn't seem to mind it. First, we put a large ball of cotton at the bottom of her feet, and her natural response was to grab hold of it. Then, with her feet comfortably supported by the cotton, we wrapped her feet with the vet wrap. Chris remarked how unusual it was for a wild raptor not to immediately use their talons to grab something in self-defense, but Shasta had never done that. Never. After wrapping her feet, we also wrapped the wing that she had a tendency to bruise, trying to keep her balance when she rode in the travel carrier. Staying behind, I said good-bye to Shasta then and wished her good luck. Chris, who had to be in Portland that evening, dropped Shasta off, saying to the vet and the techs as she left, "Take care of my girl."

My next service to Shasta would be to pick her up the following morning and take her to the clinic. Knowing that sometimes in the early mornings, my mind goes into "routine mode," I was afraid I might go straight to work rather than remember to pick her up first—so I put notes to myself all over my house as reminders and even wrote one on my hand. I doubt seriously that

I would have forgotten since I worried about Shasta whenever she went into surgery, but at least this was something I could do for Shasta until I saw her again. That night, I went to a friend's house to watch a movie. One of the women in the group happened to be a lab technician for the dentist who was working on Shasta's new beak, and I remember happily discussing her progress and what a great bird she was.

When I got home, there was an urgent phone message from Chris saying to call her back before I went to bed. Chills went down my spine as I dialed the number. When Chris answered she said, "Are you sitting down?" and I blurted out, "Shasta died." Chris said, "Yes."

Shasta had made it through the surgery fine but never came out of the anesthesia. Chris said that even the vet had been crying when he gave her this news. In her time with us, many people had come to love Shasta and were touched on a deep level by her will to survive, and this medical team had wanted very much to help her live as normal a captive life as possible.

I felt such sadness then, yet Chris and I both felt a sense of relief too, because we knew Shasta couldn't have gone through much more, no matter how kindly we all treated her and how much we wanted to help her. All night as I tried to sleep, I kept thinking this was just a bad dream, Shasta was not really gone. As morning approached, I dreaded going to work to face her empty cage. All day, trying to make myself feel better, I thought of the many animals in my life that were healthy and happy. But then I would have to tell the next volunteer coming into work about Shasta's death, and the sadness would engulf me again.

I have lost many animals to death, but losing Shasta was different somehow. For days, I didn't want to talk to anyone about it, although usually that helps ease the pain of such a loss. I realized that I missed this special bird on two levels. First, I missed her physical presence, here and now; second—well, the other level is harder to describe. What I missed was our future selves. I missed getting to spend more quality time with Shasta on walks, taking her to schools where she could have shared with others her specialness. I had thought of this future often as she'd gone through her surgeries; this was the promise of making it all worthwhile, and now the dream couldn't come true. I loved this amazing bald eagle yet never got the chance to know her as well as I felt was possible. And she never got the chance to experience more in her life beyond the medical treatments, surgeries, and recoveries.

Now, years later, the moments Shasta and I spent together in her small cage are more precious to me than ever. I am feeling she was willing to stay for a certain period of time to give us a chance to help her, but even I couldn't convince her to stay beyond that.

Fly free, my majestic one.

Chapter 3

The Coatimundi and the Jeep

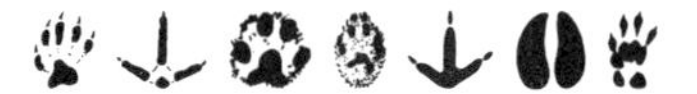

One of the interesting aspects of getting older is that you have a long trail of memories that allow you, piece by piece, to complete a present-day puzzle. This animal encounter can be traced back to my earliest memories when I was five years old and living on the island of Guam. My dad was stationed there in the army. I fell in love with three things during that year: the beautiful South Pacific beaches, the warm soothing ocean, and the old army Jeep that my dad drove. I loved the open-air rough and tumble trips we took to anywhere in that vehicle. It wasn't as luxurious as today's Wranglers or even the 1970 CJ-7. It was bare bones and loaded with personality. In my adult years, I have always wanted to own one but felt it was not a practical vehicle for my needs. Yet I still lusted after one.

After I sold my house in Florida to move to Oregon, I decided to buy a luxury item. No, I didn't buy a Jeep. But I found this unique denim jacket on the Internet: it had an embroidered cartoon of Eugene the Jeep (a character in the old *Popeye* series), riding in a rugged Jeep car. The customer could choose any color for the Jeep. I choose purple. This cartoon covered most of the back of the jacket,

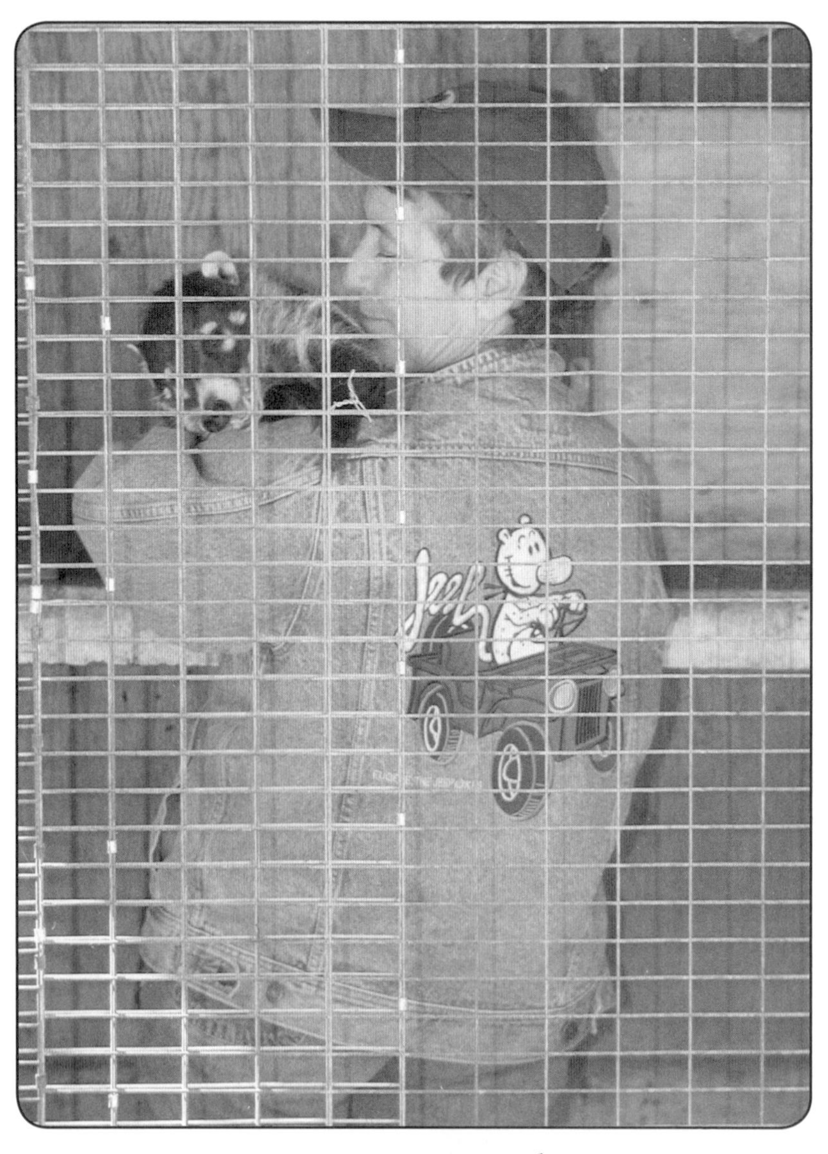

Jeep, me in my Jeep jacket

and there was a smaller version over the pocket on the front. This was my way of having a practical Jeep.

The fictitious cartoon character was patterned after an animal called a coatimundi. This is where the first piece of the puzzle fits. When I started working at the wildlife rehab center in Oregon, I met a real coatimundi. An elderly gentleman who had him as a pet could no longer care for him and gave him to the facility. "Coati" is the common name, "mundi" refers to a male. Mundi means "solitary," which is what many of the males are in the wild. Though coatimundis are members of the raccoon family, they're not native to Oregon; they live mostly in South America and in the southwest United States. Normally, our center wouldn't have accepted a non-native species, but it seemed that we were this coati's only option.

I was drawn to this coatimundi right away—probably because he had a sign on the back of his cage, near the door, warning workers not to go in with him. He had razor sharp teeth and long, powerful claws. And he was very unpredictable. Those are the animals that really need a friend, so I spent a lot of time outside his cage talking to him. After two weeks, I felt a bond and decided it would be safe to go inside. I walked in slowly and stood still in front of him. He came over to sniff and check me out. Not wanting to push my luck, I stayed only a few minutes. I made short visits like that for about a week and then let him climb into my arms. He settled in and put his arms around my neck for a hug. When an animal lets its guard down with feelings of trust, I just melt!

Each day our bond grew tighter. I began to notice how much he seemed like one of the baby gorillas I had raised many years before at a zoo. This puzzled me. Since the coatimundi was a member of the raccoon family, I would have expected him to feel more like a dog in my arms than like a human baby, as the gorillas had felt. Then I read more about coatimundis and learned that even though they are members of the raccoon family, their behavior is more like a primate's. And yet another piece from the past fell into place.

Oh, I almost forgot—this coatimundi was named "Jeep!"

My friend, Jeep

Chapter 4

Mitch: A Spiritual Experience

A man found a golden eagle with a broken wing on the side of a mountain road and stopped to help it, and then as luck would have it, another man stopped who was experienced in handling wildlife. He was able to help the first man gently pick up the eagle without injuring it—or themselves; even though its wing was broken, its beak and talons were still dangerous weapons. The first man brought the eagle to our rehab center. I found out later that the other man, the one with the experience with eagles, was someone I knew. He and his wife, Leslie, are wonderful animal advocates who work at a tiger sanctuary near Medford, Oregon. I later named this eagle Mitch, after him.

Chris (the clinic supervisor) and I x-rayed Mitch and found that his wing was broken in several places. It was fixable, but if it didn't mend perfectly, he might not be able to fly well enough to hunt. Chris wrapped the wing, and we hoped for the best. Mitch settled into his new cage, which was about six feet by five feet and ten feet high. There he had a perch to sit on and windows for sun.

I had never been around golden eagles before, so I didn't know if this one was unusually calm or if that was just their nature. He was a very good patient. We had to give him antibiotic injections twice a day, and he allowed this without a struggle. Periodically, we x-rayed him to see how the bone was healing. Most birds need to be slightly sedated before we can lay them on the x-ray table, but Mitch would lie still enough on his own. His wing was healing, but looked a bit crooked to us.

After six weeks, we removed the wrap, and a few days later, we tested his flying abilities. Since the wing bone was crooked, it appeared as if Mitch's flying days were over. Then we had to decide if he should be euthanized or kept in captivity for education programs. As hard as it is to euthanize an animal, sometimes it is for the best—especially for birds that are highly stressed in captivity and can no longer fly. Because Mitch's personality was so mellow and he didn't seem stressed among humans, we decided to keep him.

Mitch was moved to where the education birds were kept, and I didn't see him as often when he was there. At night, these birds were put in secure boxes, and during the day, they were tethered out in a yard on individual perches. They all wore jesses around their ankles, with a long leather leash attached to their bow-shaped perches. This way, they could get sun and use their wings a little. At first the arrangement looked cruel to me, and many people feel this way when they see it. But once I learned that wild birds of prey spend 80 percent of their day perched in trees and that none of these birds could survive in the wild, I didn't feel quite as bad. Again, some birds are content with this life, while others can't adjust to it. Mitch seemed to enjoy being perched outside and taken places by humans.

All went well for about three months, but then something scared him, and he quickly flew off the perch to the end of his tether. His leg took the force of the quick stop and snapped the bone, breaking it. Back to our clinic he came. Chris took him to the vet's, where he was fitted with a wire splint. The splint had to be checked every week to see if it needed to be tightened or loosened. I was sorry he was injured but happy he was back in our care. He got a lot more personal attention with us.

I continued to be amazed at the trust this eagle had in us, now that I was spending a lot more time with him again. I noticed that whenever I had to pick him up for an exam, he seemed to like to have his head stroked. So in the evenings before I went home, I held him and stroked his head until he became limp and sometimes even fell asleep. He let me just walk into his cage, pick him up, and hold him with his back against my chest as I held his one good leg. Then I would sit down with him on my lap. Few other eagles—bald or golden—would allow a human to pick them up without a chase or an attack with their talons. I knew that eagles can exert about a thousand pounds of pressure in their talons, so even though I trusted Mitch not to hurt me with his one good leg, I still held it out of harm's way. His other leg was in a cast, and he couldn't move it.

One night when I came in to check on him and give him his nightly head rub, I noticed that the wrap around the cast was coming loose. It had to be fixed by taking most of the bandage off and replacing it. But I was the only one around, and this required two people—someone needed to hold him while the other did the bandage work. Regardless, I knew this had to be done so I put him down and got materials from the clinic. When I came back to him, I held him on my lap again as usual. I knew that if he got scared or the procedure hurt him, he could easily grab me with his talons. But I decided then that I just had to trust that he wouldn't hurt me, and I let go of his good leg.

I gently began to unwrap the bandage. Once the old one was off I carefully rewrapped the splint and cut off the excess with scissors. The bandaging process took fifteen minutes, and Mitch stayed very still, watching me the whole time. When I realized what had just happened, tears rolled down my cheek and a tingle went through my body. A golden eagle had just blended with me in trust. I wonder if he knew what an incredible gift that was to me.

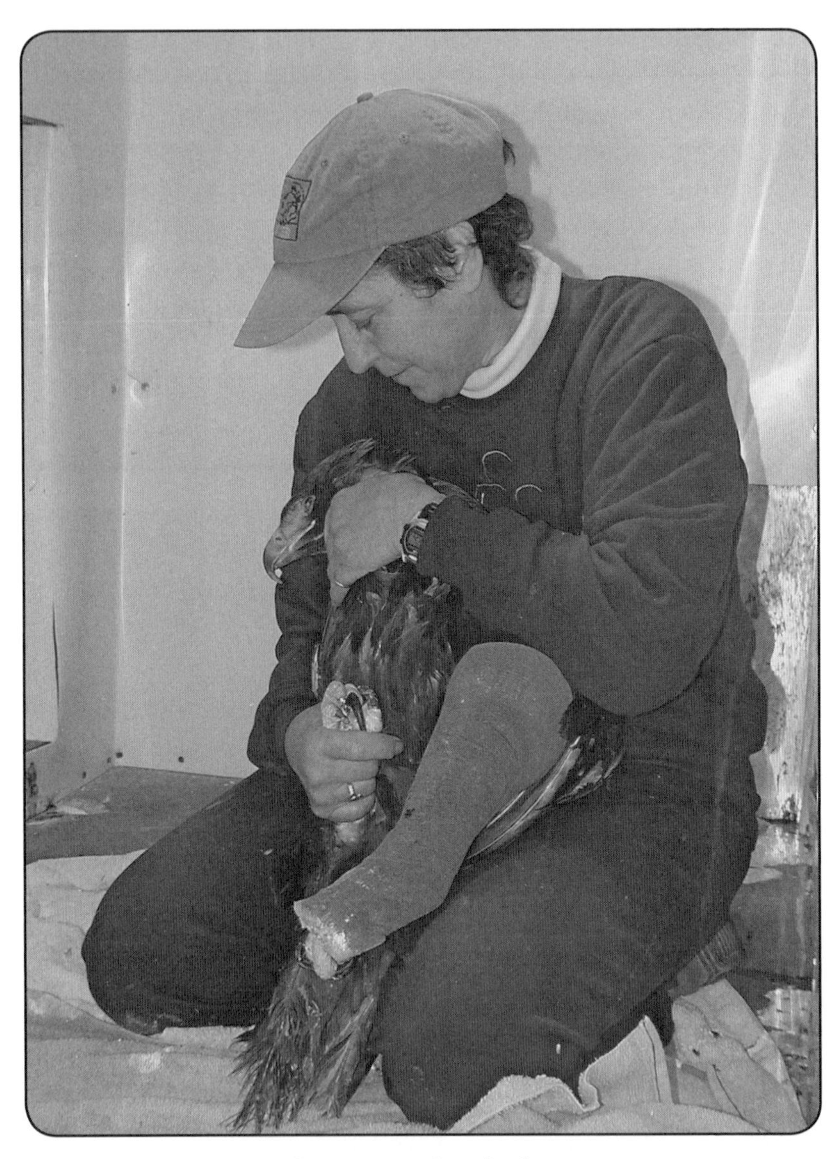

Calming Mitch at bedtime

Mitch's leg continued to heal, but it took more time than usual. In about eight weeks, we took an x-ray, and though it was a little hard to tell from the angle we had, it looked as if it had healed. We removed the cast and put him in a large mew (a cage for birds of prey). He had two long perches about five feet apart so he could fly that short distance back and forth.

All went well for another three months, but then we noticed that the previously broken leg seemed to be bending. When the bend became more noticeable, we took an x-ray, and it looked as if it was bending at the old break point. We had to take an x-ray picture from a different angle to be sure, and this would put him in an uncomfortable position, so we sedated him. From this view, we saw that the pieces of broken bone had healed side-by-side instead of end-to-end. The only way to fix this was with surgery. The vet would have to cut the bone apart and then cut off the tips of each so they could have a fresh chance to mend with each other. There would be no guarantee of success, but management decided to take the chance.

It was a difficult operation, taking five instead of the expected three hours. The vet had to cut delicately through muscle to get to the bone, careful not to cut tendons. Then he had to saw the bones apart. Cutting the tips off the bones was a bit harder than expected. He had to try several types of equipment. Next he had to put a pin down the center of the bone. I watched the whole process, finding it fascinating yet a bit disturbing. It was very bloody, and I winced many times when it looked as if Mitch was being hurt, even though he was peacefully sedated.

When the operation was over the incision was sutured together and a new cast put on. We wrapped Mitch in a blanket (he was still slightly sedated), put him in a pet carrier, and drove him back to the center. I laid him on the blanket in his cage and watched as he slowly came out of his stupor. After only two hours, he was standing, albeit a bit unsteadily, as he got used to having his peg leg again! Then he was back on the bumpy road to recovery.

But Mitch's bone healed very slowly, and we were beginning to wonder if he had a genetic bone problem; maybe that was why his bones never healed as quickly as the vets expected. We took x-rays every two weeks, and the bone was grafting but not as well as it should have. Finally, two weeks later than the normal four weeks of healing, the pin was pulled out so the bone could continue to graft without including the pin. We noticed in the x-rays that the bone continued to have slight space between the grafting points. It just wasn't healing properly.

Then one day when I was at home on my day off I got a call from the other clinic tech telling me she thought they were going to euthanize Mitch. I rushed in to plead for his life. The director knew how much Mitch meant to me so he said I could take one more x-ray, but if it hadn't healed yet, it wasn't going to, and Mitch needed to be euthanized because he couldn't live with one leg. The x-ray revealed the bad news. He would have to be euthanized. I couldn't give the injection myself, but I did hold him in my arms as a vet tech gave him the drug.

After Mitch was gone, everyone left the room so I could be alone with my friend. I held him close to me as I had done many times in the past, only this time he was asleep forever. I kept thinking I had let him down, that I should have fought more for his life or come up with an idea of how to make an artificial leg. The guilt was overwhelming. The tears wouldn't stop flowing. Oh, how I wished I could bring him back. Yet knew I had to shake the guilt; it is such an unhealthy emotion.

I thought to myself that Mitch would have died a cold, lonely, painful death if someone hadn't found him on the side of the road. I gave him a year and a half of love and good care. He died peacefully. For the next few weeks, I worked on getting rid of the guilt by feeling the rare gift of love and trust this eagle had given me and knowing that I'd done the best I could for him. I knew Mitch only for a short while, but he would remain my treasured friend forever.

Chapter 5

Little Guys Escape!

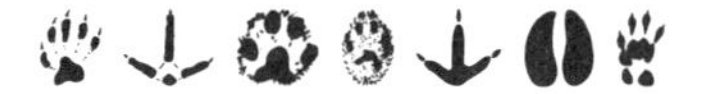

Given the human element in the long-term care of captive animals, certain troublesome surprises invariably happen. Escapes. Luckily, many are minor escapes, such as when a volunteer leaves a door unlatched after cleaning the cage, of our permanent animals. "Minor" because they involve semi-tame animals used to humans.

Ruby the raccoon, one of our long-time residents, opened her unlatched door one day when no one was around. Once it was discovered she was missing, we all went on the search. She was spotted about fifty yards away digging in the nearby creek. I ran to our produce room and grabbed a cup of grapes, her favorite food at the time, then walked calmly near her. I made a trail with them, leading to her cage, grapes about a foot apart. Ruby followed them as she ate, just like Hansel and Gretel. She would pick one up with her delicate paws and roll it around her palms before plopping it in her mouth. Raccoons' paws and mouths have many sensors, making them very tactile. When Ruby placed the grape on her tongue, she put her head back and let the juices drip down her throat as she chewed. Quite the slow process, so it took a while for her to make her way to the makeshift step I put by her door so she could easily step into her cage that was four feet off the ground. Other than the time it took, there was no problem.

One of the funnier accidental releases was again when a door was left unlatched. I was working in the food prep room and looked out a window to the food-sorting room. I saw the door to the outside opening but could see no person pushing it. I was thinking it had to be a small person to open the door without being seen. I couldn't think of any kids who were there that day so I went out to investigate—and found an otter shuffling around on the floor. It took me a few seconds to realize it was Zoey our otter and not a visitor from our nearby creek. I found it interesting that once she was out of her cage she could have gotten in the creek just below her cage and followed it either to the river in one direction or to our large pond in the other, but she had chosen, instead, to go to her friends, the people who feed her.

Zoey wasn't as easy to get back to her home as Ruby. The food she likes the best is fish. Not only did I not have enough fish to make a trail, but as far as she had to go, she would have been full by the time she was halfway there. Besides, I think she prefers people more than food so she may not have fallen for that trick. I got a large crate, picked her up by her tail, and put her in. She likes to bite in play so I didn't want to chance those playful sharp teeth to make contact with any part of my body. She weighed about thirty pounds so a volunteer took one end of the crate and I took the other. We put the crate on the golf cart and drove her to her enclosure. I can't say she was really happy to return, but she was safe, and we were safe.

Escapes come in many forms. None are planned, and all provide unforgettable memories. This one can't really be called an escape because we were releasing the animal. I suppose it could be called a "planned escape."

We were releasing two skunks and six raccoons in a wilderness area near the Rogue River. When at all possible, we like to put the animals back near where they were found. This way they will recognize the territory and perhaps have a home base where others don't intrude. But many times, the animals have been hit by cars or are found near homes that could endanger them or the humans. Then we just have to pick a spot where that species would normally live and hope for the best. Even in the best of circumstances, life in thc wild isn't casy.

We drove in on the maintained road as far as we could and then had to carry the three dog carriers to the off-road areas. One person could handle the carrier with the two skunks, but four had to take the two other carriers with three raccoons in each. We had to go through a wooded area first before walking down a hill to the river. We decided to release the skunks near the brambles in the wooded area so they would have shelter and be near the woods and water. We set the carrier on the ground and opened the door. One ran out to the blackberry bushes. In a few minutes the other peeked out, took a few quick sniffs as it surveyed the area, decided it was safe, and meandered off to the bushes. When most animals are released, they either take off in a flash like the first skunk or take their time like this last one.

Occasionally, there are exceptions when an animal feels more secure in the small protective carrier than the huge unknown of the outdoors. That is what living in a cage for several months as an orphan can do. In these cases, the carrier has to be tipped up a bit so the animal slides out. The slide to the ground sometimes kicks in the "I want to be free" instinct, and they are off.

However, a tiny minority of even more hardcore security types (like me, not wanting to abandon my secure safe room when it was time for me to move away from home) require more urging. An animal like this won't be encouraged by just a slight tilt. It holds on to the side vents of the cage so it won't slide down even when the cage is tilted on its end. This is when several sound thumps on the ground with the carrier are needed as well as some slaps on the

An orphan skunk just before his "hands off" stage to prepare for release.

sides. It is not the nicest way to force the transition from security to "You're on your own, buckoo," but it has to be done.

Once the skunks were out and about, we trekked on down to the river. We put the two carriers with the raccoons down at the water's edge. In hindsight, it might have been better not to have placed them so close to the water. When the doors were opened, they all took off like a shot along the bank or up towards the woods, except the smallest. It went straight in the water. He was so scared; I don't think he even knew it was water until his feet suddenly weren't on solid ground and he was already six feet from shore. I was sure he would drown and expressed my fear to the others. This was my first raccoon release, and the others had done it many times. They assured me that raccoons could swim, and it would be fine. I trusted their expertise and let out a sigh of relief.

But as I watched this little guy's tennis-ball-sized head bob as he struggled to get back to shore, I noticed he was staying in one place as he paddled and his head was sinking. He was caught in a current. My brain quickly processed that the experts were wrong. The raccoon was drowning. As I jumped in the river, others yelled something about danger, but being from Florida where rivers don't have such strong currents, it didn't compute that I could be lost too. Luckily it was only neck-deep for me where the raccoon was. As I approached, the little guy seemed to sense that I could help, and he tried to swim toward me. When he got within arm's reach, I grabbed him. I had already started back when he discovered I was a human and to be feared. He tried to bite me so I had to toss him toward the shore as I swam. After several tosses, I grabbed him one last time and tossed him once more toward the shore where he fell on dry land. As his soggy little feet hit solid ground, they took him quickly in the correct direction, toward the woods.

I hadn't lived in the area very long and had always wanted to experience the Rogue River. Now I had.

Chapter 6

When Big Guys Escape

Bears!

The first baby bears that came into the center's care were two months old and weighed about four pounds each. I don't remember why or where we got them, just that they were the focal point of everyone's interest at that time. A local hospital had recently donated two incubators to the center, and one of these became the bears' new warm nesting place.

They were twin brothers with very different personalities. One was relaxed and easily fed with a bottle while the other seemed less secure and was a very fussy drinker. New at raising bears, we were told to follow some pretty strict guidelines set down by the Department of Fish and Wildlife. One of the rules was that we were not to name them, so (still needing to identify them in some way) we labeled them "B1" and "B2." Of course these functioned as names as well; what difference does it make to bears if their names are spelled "B1" and "B2" or "Beewan" and Beetu"? But the distinction satisfied the law, so everyone was happy (including Beewan and Beetu).

A few days after they arrived, we noticed that they were losing their fur. A skin test at the vet's diagnosed the condition as ringworm, so we had to give the boys medicated baths once a day. This eventually got rid of the problem, but they were a motley pair for quite a while and got the unofficial nicknames "Barnum and Bailey, the ringworm twins."

The Department of Fish and Wildlife's policy on naming was based on the belief that a less handled/nurtured bear would be less likely to grow up to become a problem bear in the wild. Later, when I did some research into that theory, my hunch was confirmed that they were wrong about that. All animals need nurturing to be psychologically healthy, just as the human animal does. All the other animals we raised as babies at the center were nurtured—but once they were weaned, they were put on their own and quickly became the wild creatures they were meant to be. Of course, if bears are handled and tamed as they grow up *after* being weaned, that's another story.

In any case, for the first month, we strictly followed the department guidelines and held the bears as little as possible—just during their bottle feedings. It helped to know that at least they had each other to cuddle with. But as they developed, it was obvious that they were becoming neurotic and insecure, crying out for attention whenever they saw humans. I started holding them more, and that seemed to calm them down.

In my research, I found an extensive survey of many places that had raised and released bears and the widely varying rearing techniques that had been used. Some followed a strictly hands-off technique, with humans dressing in bear suits and never letting bears hear their voices; some cuddled the bears, even letting them sleep in the caregivers' beds; and there were other methods in between these two extremes. The results? It seemed to make no difference; the percentage of problem bears was low in all cases. I also found research confirming that both nurtured animals and humans are more secure as adults. It seemed to me that the more

Chris gives B2 his bath

secure bears would be less likely to attack people, at least out of fear. So I decided to use the more nurturing technique.

At about five months of age, the bears were fully weaned and we moved them to a nice, large outdoor enclosure. It was about thirty feet by thirty feet and had trees to climb and logs to play on, as well as a cozy wooden den for shelter. The bears were making good progress, developing all their bear survival skills, and we helped by scattering their fruits and veggies all over their habitat to encourage them to forage for their food in a way similar to what they would be doing once they were on their own in the wild. I'm not sure how much they weighed at this point; they were too big and uncontrollable to be put on a scale by then. But when they stood on their back legs they were about three feet tall and quite strong.

When they were six months old, they were almost four feet tall and too strong and dangerous for me to go into their enclosure any more. I remember thinking that when I had worked with gorillas, I could still go in with them when they were six years old, but I couldn't safely go in with bears at six months! The main reason for our caution was that bears have very sharp claws and teeth that they use in play as well as when they're highly motivated by food. If they think a person has food, or if they even smell food on a person, they can become very aggressive trying to get it. Gorillas couldn't care less! It was just as well, though. Since Beewan and Beetu were weaned, it was time for them to be bears and not have human attention any more.

About three weeks after they had been moved into their large new enclosure, little Beetu, the more introverted one, somehow got loose, and an employee spotted him about thirty feet up a tree. The bears' enclosure had a ten-foot-high chain-link fence with another four feet of thinner wire fencing leaning inward at a sixty-degree angle. This design was intended to prevent bears from climbing out. We weren't sure how Beetu had gotten out but thought it was possible he'd jumped over one of the metal baffles. Surrounding the trunk of each tree in the bears' enclosure, the baffles supposedly

Giving B1 & B2 the nurturing they deserved

prevented the bears from climbing too high. But if he'd gotten past one of those on one of the smaller trees near the holding enclosure, he might have climbed high enough to bend a small tree down and jump onto the enclosure. We used the structure as a secure place to put the bears while we safely went into the big enclosure to spread their food and do necessary cleaning. This smaller enclosure had four sides and a top of chain link with a door to their larger pen. Regardless of how he had done it, Beetu was now near the top of the large tree growing inside this enclosure; it went up through the top for another twenty feet. This put him nearly thirty feet from the ground (the top of this cage was built around the tree that grew through it).

Chris and I were called to help with the situation, along with several of our maintenance men. Since I was one of the bears' "moms" and had once been trained by an orangutan (I worked with 20 years earlier) to climb trees, I offered to climb up to coax Beetu down. When I got close enough, I hugged him and he leaned over to lick my face. He was scared to find himself in this fix and just needed reassurance. I climbed back down, staying just out of his tongue range so he had to keep climbing down in his attempt to lick me. We continued this procedure until I got him down to the top of the smaller enclosure.

Now he was out of the tree, but we still needed to get him back into his big enclosure. Meanwhile, his twin, Beewun, was becoming nervous, and we thought he could become a problem if someone went into the enclosure with him to try to coax Beetu down. So we put some food in his night den. That was a five-by-five box set on a platform four feet off the ground. The front door of this little house was inside their enclosure, and the back half was on the other side of the fence, making it possible for a person to close the door with a sliding handle while standing safely outside the enclosure. When Beewun went in to eat the unexpected snack we put there for him, we closed his door. Once he was out of the way, several of us went inside the bears' main enclosure to use food to entice his brother to come down. But Beetu was still frightened, and instead of coming down, he decided to climb onto the fencing that leaned inward; then he walked about ten feet around to the night den that held his brother. He nearly jumped onto the porch of the den, but then he saw someone he didn't know and turned around to go back toward the tree he'd climbed. I waved my hands and yelled a stern NO, and amazingly, he went no further.

Finally, after about forty-five minutes of this back-and-forth drama, someone inside the enclosure bent a small tree up to the fencing Beetu was traveling on, and he climbed down that to the ground. We enticed him into the smaller holding pen (whose tree he had climbed) and closed the door. Now the maintenance crew could go in and fix and tree guards and cut down any "escape trees" that had made the enclosure less secure. Soon, both bears were reunited in their large enclosure, excitedly gorging on a pile of grapes to celebrate a happy ending to the unusual day.

About two weeks later, Beetu once again demonstrated his talent for being the escape artist. This time he had gotten halfway up the fence but was stuck between two small "hot wires" that had been put up after the last escape. One wire was about two feet from the ground and the other about five feet up from that one. Even though we turned the electricity off, he was still frightened of the wire and wouldn't back down. But he wouldn't go up either, and that was good, for there was a chance he could have gotten out if he had!

The maintenance crew suggested pulling Beetu off the fence using a lasso around his neck, cowboy-style, and they tried that. But they were no match for his strength. He held on for dear life and they couldn't budge him and, on top of that, they'd made him angry, and he began to growl and cry. I protested their tactics and, when they refused to give up I left in a huff and went back to the clinic. Within ten minutes, I got a call to come back to see if I had a better way. I wasn't sure I could do better, but I could certainly "do kinder." I quickly scavenged some food from our food prep area and went back to Beetu. I asked everyone to leave me alone with him because I felt that the three strangers probably only added to his fear and confusion.

After they left, I found a good-sized log and put it near him so that it was over the lower hot wire to the ground. Then I brought out my arsenal of food. I showed him a pack of hot dogs (a favorite) and a bag of cookies and chips (goodies donated by a local grocery store). He stopped his fussing and looked as if he might get onto the log and climb down for these tasty morsels. He put his front paws on the log and started to put his back legs on it when something scared him again. Back he went, up the fence and crying. Next in my stash was something a bear would go through fire for: a piece of iced birthday cake. He definitely wanted that and not only put all fours on the log but started down. When he was halfway down though, the log tilted, and he almost fell—but instead of going all the way down with it, he jumped back to the fence. Now he no longer trusted the log, even for a sugary reward.

I reached over and stroked Beetu's head, trying to calm him down. After a few minutes he stopped fussing but he was still upset. I was thinking of what to do next when I remembered that our maintenance department had several ladders stored not too far away. I left the enclosure and walked about fifty yards to their building. When I got there, I found only two. One was about ten feet long and very heavy; the other was even longer. Both were bright orange. At first I was afraid the color would scare the little guy, but felt I had no choice. I lugged the smaller of the two back to the enclosure and put it up next to the fussy growling bear.

Just as I put the ladder up, I got a call on my radio from my supervisor, asking how things were going. I answered, "Okay!" But Beetu's growling and crying in the background made my supervisor a little nervous. She didn't want me up there alone in case the bear got really scared and jumped on me. I have to agree that I probably wouldn't have come through that situation unscathed. She said she was sending the maintenance people back.

I knew I had to get the bear down before the cowboys arrived, fearing what they would try next! I put the cake at the bottom of the ladder and then went out of the enclosure and hid behind a tree in case Beetu just needed to be alone for a while. A couple of minutes after I disappeared from view, he stopped crying and looked at the cake. Then as normal as could be, he climbed down the rungs as if he had been climbing ladders for years! Then I went back in and tossed some of the hot dogs to the far end of the enclosure. When he went after them, I retrieved the ladder and left. I stepped out and closed the door just as the maintenance crew arrived. Lucky bear!

They were released at about 1 ½ years of age....they never became problem bears!

Bobcat!

This last escape story, starring Rufus, a young bobcat, melted my heart. I had known him since he was two months old. Three young boys on a hike had found him and his two littermates in a den; the mother bobcat had apparently gone off hunting. The babies were just about five weeks old then. Each boy took a cub home. The parents of two of the boys told them to take the tiny cubs right back to their mother, but the other boy got to keep his. That is, until the authorities found out about the stolen cat three weeks later. The parents were fined, and the kitten, by now very tame, was brought to our center. We named him Rufus after the scientific name for bobcats. By that time, Rufus was now far too "humanized" to be released.

I think he was destined to be in captivity. I have seen some animals from the wild adjust only reluctantly to captivity when they can't be released, but Rufus loved his human companions and his safe, secure cage from the start. He had a house-cat personality—with a few wild edges! The first month he was with us, he was allowed to play with us (staff and volunteers) in the clinic during the day. But soon, he became just a bit too wild in his play to be safe with everyone, so after that he was taken for walks every day to a large play yard for supervised exercise and play with handlers experienced with bobcats.

When baby season hits at a rehab center, staff time is stretched to the max with no longer extra time for things like unscheduled playtimes with a young bobcat. Now Rufus had to stay in his cage most of the time, getting only short bouts of attention during cleaning and feeding. He seemed to be content with this as he had a lynx in the cage next to him, and they played with each other through the fencing throughout the day. I think he knew he was loved and had quality time when the quantity of time allowed it. He was still a happy cat.

Since I didn't have time during the day to be with him, I always took some time after work, no matter how late it was, to go in with him for some one-on-one quiet time. When he would see me approach his cage early in the evening, he knew what time it was and would go straight to his bed of straw and lie down. He knew I would soon be curled next to him, lying catlike, with one arm under his head and the other petting his head and chest. He would put his paw around my neck and purr until he fell asleep and the purr went silent. I might have spent only twenty to thirty minutes with him like this, but doing it every evening I was there created a close bond between us that became very important to his safety during his escape adventure a year later.

Rufus was about a year and a half old and a full-grown bobcat when a professional photographer, who was to be a friend of the director, asked if he could get some photos of Rufus in a wild-looking setting. The director gave permission for this and asked me to take Rufus to our cougar enclosure. This was a fenced-in, natural wooded area off the cougar cages. Since cougars aren't social and don't get along with each other, they were put out in the huge area one at a time each day. On the day of the photography shoot, the cougars were all locked up in their individual cages so I could bring Rufus in to play in their yard. It was a great place to take pictures—a photo-op with grassy hills and large boulders among

Rufus playing in our break room

many pine trees. At first, I tried walking Rufus from his cage to the cougar yard on a leash but soon realized that could take a year with all the distractions along the way for Rufus to investigate, so I gave up on that and brought him there in a large carrier on the back of a golf cart.

Once we were all inside the enclosure with the door closed, I let Rufus out. He was excited about this new territory. His nose was intrigued with new scents—mostly from cougars, I assume. As he was sniffing close to the edge of the cages where the cougars were kept, he looked up and saw a huge cat stalking him at the fence's edge. It was confined in its cage, but Rufus didn't know that. In the blink of an eye he took off across the exhibit as far away from the cougars as possible! And he didn't stop at the fence barrier. He went straight up the fence. This enclosure, like the bears', had a fence with the top leaning inward, and also two layers of hot wire at the top to discourage the cougars from getting out. Somehow Rufus was able to climb between the hot wires to get on top of the leaning fence. I ran out of the enclosure to get to the other side to encourage him to come down since I figured there wasn't much chance of getting him to come back through the hot wire on the inside. My coaxing didn't work. He was too frightened and wasn't willing to jump twelve feet to the ground. He did turn around to face me and gave me a "help me" look, but when he did that, he brushed up against the hot wire. This scared him even more, and he zipped straight up a large tree next to the fence. Now he was another ten feet up, and from this vantage point, he spotted the bears in their enclosure about a hundred yards away. This further increased his fright level!

The vet tech at the time, Koren, and our animal trainer, Larry, came to help. They put a ladder against the tree so I could climb to reach the first of the branches. Larry handed me a long leash and collar and told me to put it on Rufus so we could gently pull him down from the ground. Knowing Rufus the way I did, I knew that wouldn't work, but I tried anyway. I climbed up the branches to

Rufus and put the collar around his neck. Larry then gently pulled the other end. Rufus just pulled backwards instead of coming down. I was afraid this might cause him to fall backwards and possibly get his leash caught on a branch and hang himself. So I nixed trying that procedure again.

Next I got some of his favorite food, chicken, and showed it to him. It didn't break his concentration on the bears, so I climbed up the tree to get the treat closer to him. Once he sniffed the chicken, it did grab his attention. He moved a paw as if he was going to come toward the meat, and I started to climb down, hoping he would follow. He followed me briefly, but with his eyes only. Soon he was fixated again on the bears. Since the bears seemed to be a distraction and a hindrance to getting him down, Koren and Larry took off in the golf cart with a bucket of food and fed the bears in a place that would put them out of Rufus's view.

That did settle him down, and I made my tenth trip up the tree to comfort him and try to encourage him to come down. Each time I would put my arms around him and stroke him, he would relax and purr, but when I would try to push him off his branch to encourage him to follow me down, he would just hold on tighter. In the middle of all this, I remember hearing the photographer saying something like, "Could you move aside so I can get just a picture of the bobcat?" He made sure he got a few pictures before he left us to our dilemma. I would love to see some of those photos, but I never saw him or his pictures again!

After about an hour of this, it began to rain, and I knew I had to do something quickly or I'd never make it down safely, with or without a bobcat. I finally told him, "Rufus, you have to trust me—*please*." Then as I held his forty-something-pound body, he held onto me for dear life, his front paws over my shoulder and digging in with his claws. By now my legs were like rubber after climbing that tree so many times.

As I slowly lowered us down through twenty feet of now rain-slicked branches, I wondered where Koren and Larry were. I would

Rufus and I posing for my holiday photo

need the crate for Rufus once we got to the ground and especially would have liked some help getting down the ladder. At last we were past the branches and at the top of the ladder, and here came Koren and Larry in the golf cart. Larry was tall enough to help me lower Rufus from the ladder. When Rufus's feet touched the ground, he just wanted to run, but I was able to get down in time to hold and comfort him until he could go safely inside the crate. Once I knew he was okay and the ordeal was finally over, I collapsed on top of the crate and released all my tension with a brief cry—and then a big smile.

On our drive back, I heard Larry and Koren talking to me, but I was in a fog-like state of mind. All I could really feel was a mystical connection with Rufus—a soul connection forged in trust that only I, and perhaps Rufus, could understand.

CHAPTER 7

SQUIRRELS IN MY LIFE

In my first job, I raised baby animals in a zoo nursery for seven years, and in that time I was a mom to leopards, lions, tigers, and—yes—bears, but my favorites were seven baby gorillas. Gorillas became my main focus, in fact. I studied them, read about them, and visited them; they were my heart and soul. Thirty years after that first zoo experience, I began working at a wildlife rehabilitation center where, instead of lions, tigers, and gorillas, I was caring for raccoons, bears, and squirrels.

I don't know what it is about squirrels, but I just love them.

I began working at the clinic in August, the end of the "baby season." I'd heard a lot about baby season and how exhausting it was, but until I experienced it the following year, I had no real concept what that meant. Until then, I thought my work was exhausting; taking in sick and injured owls, foxes, skunks, and hawks was always an intense adventure. I would go home at night thinking and worrying about the day's patients. Most work days didn't end after eight hours—it was more like nine, ten, or eleven, so exhaustion was the norm … or so I thought. But when February rolled around, and the baby season started up again, it was nonstop babies from then until late September to early October.

My first orphan baby animals at the center were squirrels. There were three; the siblings had been blown from their nest during a nighttime storm. From birth to about two weeks of age, little ones like this are called "pinkies." They are very small, about the size of an adult human thumb. Where the eyes will be, they have little bumps. Their ears, not yet opened, still lie flat against their heads. They are completely furless and helpless. Since I had never raised a baby squirrel, my boss, Chris (nicknamed "Dr. Chris"), who had eight years of experience, took these home the first two nights. This is a critical period, for if everything is not done correctly, the pinkies can easily die. Of course, any animals arriving at our rehab center already had a strike against them because some kind of trauma had brought them to us. A rehabber never knows what internal damage has been done when a baby this young falls from a nest. And feeding something so small takes both patience and control. We used a tiny rubber nipple secured to a 1 cc syringe (minus the needle). Most other newly acquired babies struggle with a strange nipple that isn't mom's, but baby squirrels usually adapt quite quickly. The hard part is to push on the syringe just enough to help the baby suck the right amount. If there's too much liquid, it can get into the lungs, causing pneumonia, which can in end in death.

After the first two nights of special care, Chris handed the pinkies over to me. What a miracle it was to watch them change daily. Their little bodies gradually turned from pink to grey as their fur grew in, little slits appeared on their "eye bumps," and then their tiny ears popped up, their eyes opened, and their tails became fluffy. In a matter of just eight weeks, they turned into miniature squirrels! They sat up and held nuts in their little front paws as their tails curled up over their heads….just like the adults squirrels you see in your yard.

With these three babies, the squirrel species crawled right into my heart and knocked the gorilla to second place. I'm sure that for some people, the idea of a tiny squirrel replacing the magnificent gorilla might seem a bit puzzling. Some say perhaps I was a squirrel in my past life or maybe a tree! Others just say I was, and still am, a nut.

All squirrel babies open their eyes at about four weeks of age, and at that stage, we offer them solid food for the first time—usually a small piece of apple or a piece of grape with the skin removed in case they might choke on it—in addition to the formula. At about eight weeks, the weaning process begins, and by twelve to twenty weeks, they're off the formula completely. At that point, we put the youngsters into a pre-release cage to get them ready for release, and we lessen the amount of human contact so they can become who they were meant to be. Some squirrels "wild up" more quickly than others, but most do it rather rapidly.

Peanut

As I prepared my second batch of three squirrel orphans for their pre-release cage outdoors, I decorated it with naturalistic materials—tree branches, dirt and pine needles for a floor, and a nest box, which I attached to the wall about five feet off the ground. I put some cuddly blankets in the box, too, for it was a bit colder at night outside than what they were used to. The purpose of putting them outside was to get them used to the natural temperatures they would soon be experiencing as free, adult squirrels.

It was always traumatic for me to put the orphans outside for the first time. In rehabbing, as cute as the babies are, we are not supposed to name them. Naming automatically forms a psychological bond, and that makes it harder for the rehabber to release them when the time comes. Still, as with every rule, there are always exceptions (I'll tell you about some of mine later on). But even though I didn't name them, I still became attached. The pre-release cage, about six feet by four, was big enough for me to go in with them, and I did; I stayed for a while and let them climb on me as they got used to this new home. Once they seemed settled

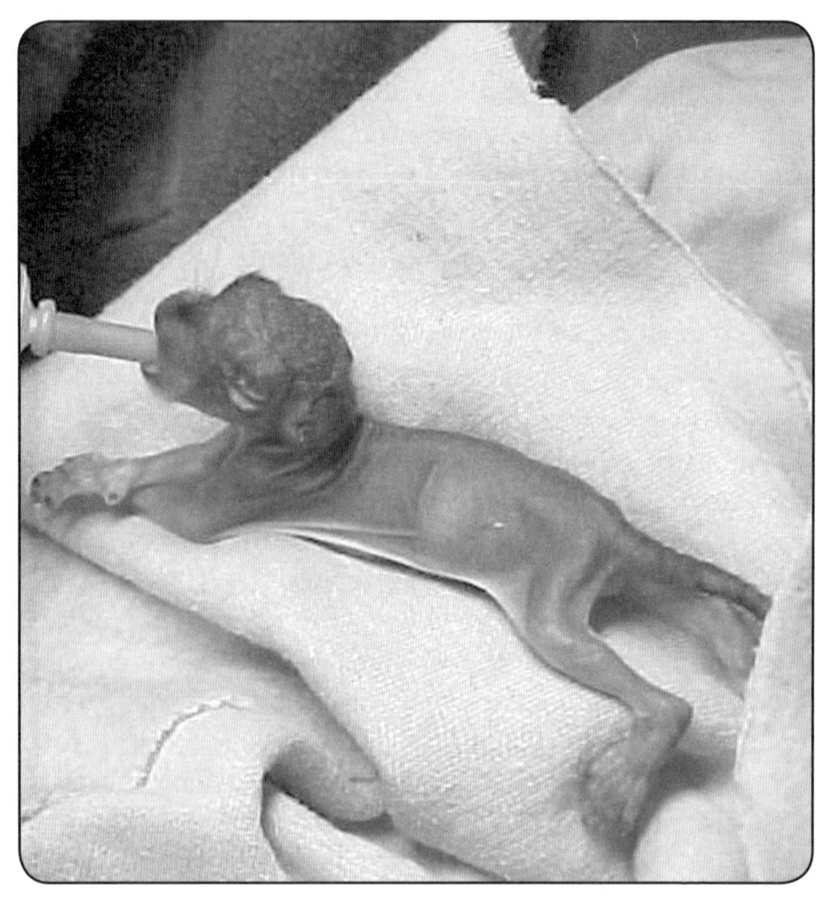

A “pinkie” squirrel being fed

into the nest box, I left, but throughout the day, I checked on them to make sure they were okay. And they were. They seemed to be having a great time, climbing the tree branches and playing in the dirt. They were beginning to learn some of the skills they'd need for a wild life.

That night, as on all first nights, I worried how they were doing and lay awake for hours. No good night's rest for me; I got up with the first light and quickly got ready for work. After parking my car I headed straight for their cage to ease my worry by making sure they were fine. But as I approached the cage, I saw a small grey furry bundle on its floor. I didn't think they would be up so early, so I went in to see why this one was there. He saw me, and my heart sank as he tried to come to me but couldn't use his back legs. I picked him up and his hind end was limp. I assumed he had fallen and broken his back. I rushed him into the clinic to show him to Dr. Chris soon as she arrived and we looked the small animal over. My eyes teared up as I feared his condition would render him helpless and he would have to be euthanized. But Chris gave me hope; she said sometimes, if the animal is kept in a small area for about a month, this kind of injury can heal.

So I put him into the small cat carrier which would be his home for the next four weeks, and I took him home with me every night. He had been used to siblings and attention so I took him out every so often to hold him and lovingly talked to him to encourage the healing process. I didn't want to name him because I felt he would get better and be released, and I was attached enough as it was. I wanted to call him something, though, so I thought "Peanut" would be a nice generic name that wouldn't be as endearing and attachment-forming as perhaps Rocky or Elmo (how irrational is that?).

A month went by without much improvement in the patient. At work, we had so many animals coming and going that this one was pretty much forgotten ... not by me, though. I continued to keep him at home for fear he would be put down if they knew he still couldn't use his back legs. At the back of the property at the center, there was a huge junk pile with many objects that might be re-used, and there I found an old cage that would give the squirrel more space to crawl around. By now he could actually get around very well with just his front legs. I put this cage on the floor in my computer room at home, and let him out when I was there during the day. He was pretty amazing—he scooted all over the house, up and down furniture and curtains. He even could zigzag around my four cats, which had no interest in him as they might for an outdoor squirrel; I guess they knew he was a part of the family.

Since I'd never had a squirrel as a pet, nor would I promote such an idea, I wasn't sure how to handle him. Should I encourage him to be a wild squirrel and not touch or pet him, or should I try to tame him? As it turned out, he tamed himself. I just let him come to me when he wanted, and he started spending more and more time on my lap. Without thinking, I'd pet him; at first he was a little uncomfortable with this, but soon he loved it, which made it hard for me to type on the computer. Before long, he was no longer content to sit on my lap but began demanding to be petted. In fact, if I stopped, he would run all over my body until I did pet him. Whether he was running or being petted, computing became very difficult for me, but I thought it was a small adjustment for having this little squirrel in my life.

But as time went by, though his spirits were high, Peanut always seemed sickly. His fur was dull and rather sparse. And occasionally there was blood in his urine. I realized he must have had other abnormalities internally, perhaps from his injury. Then, when he was about six months old, I noticed something pink trailing him as he scooted around on the floor. I picked him up and saw that his penis had prolapsed. Half of the inside was now outside his body. My first thought was that this was the end; I was going to lose him. Who could fix such a problem, and how could he survive if they couldn't? I rushed him in to work, and Dr. Chris took him to our vet. She also had to take a raccoon to have a surgical pin removed

Peanut disrupting my computer work

from his leg after surgery had healed the broken bone. She dropped both animals off at the vet's office. The staff said they would call when they had finished. I told Chris to tell the doctor that if Peanut couldn't be fixed to euthanize him there so we wouldn't have to do it.

I tried to keep busy with chores for all my other animals so I wouldn't worry. Finally at four o'clock, the phone rang. One of the doctor's assistants was on the line and said, "You can come and pick up your critters." The word *critters* transformed my fear to joy, but to make sure this wasn't a mistake, I asked, "Both of them?" She said, "Yes." I couldn't contain my tears of joy. A volunteer who knew how much I loved Peanut was standing nearby, and she couldn't hold back her tears either—and she was someone who didn't shed tears easily.

I was still puzzled at my deep love of squirrels, but I would count every day with little Peanut as a blessing.

And so my life with disabled squirrels began. I became the rehab center's unofficial queen of the special cases. Most wildlife rehabbers would euthanize disabled squirrels. I knew, though the chances were slim, some animals could reverse the disability and fully recover, then be released. They do not make good pets for a number of reasons. They only bond with their main caregiver and usually are aggressive toward anyone else. They are a high maintenance and expensive animal to keep. They need large cages, lots of good, expensive food and supplements, special lighting, and much more. Not to mention the fact that, in most states, it is illegal to have them as pets.

Peanut, my first special squirrel, didn't stay with me long. Three months after that emergency trip to the vet, when I thought I'd lost him, I really did. He accidentally escaped the house when I thought he was asleep in his cage. While I coaxed a newly acquired kitten to come inside, he slipped out unseen. When I heard a splashing noise coming from the pond in the back yard, I ran to the back window to see what was happening and saw an animal in the water. I tried to open the window so I could climb out to get there faster, but it was stuck closed. I quickly ran out the front door and around to the back, and grabbed the animal just as he was going under. Only then did I realize it was Peanut! His little heart was still beating, but his life was already slipping away. I tried to resuscitate him by blowing air into his mouth, but mouth-to-mouth with a squirrel is not easy. His heart stopped—and my life seemed to stop with it for a few seconds. The grief was overpowering. I almost fainted. My little squirrel was gone.

Losing animal children has been a constant in my life. I have learned to survive, but it takes a toll on my heart. I think when my time is up in this life, it will be when the holes in my heart make it impossible to beat any longer. Fortunately, I do have many joys to patch them over, as life goes along. Many years have passed since I lost Peanut, but there is still a tear in my soul for him.

Monkey Boy, Silver, and Chrissie

I raised many squirrels in the five years I worked at the center. Most that came in during our hectic baby season didn't have names; they grew, and they were released. There really wasn't time to get to know them, and there were many other animals to care for, as well, between February and October.

The special ones were the ones with some kind of disability. Monkey Boy climbed around like a monkey and needed special care between the ages of four and nine weeks because he had some internal problems that prevented him from defecating properly. We had to add mineral oil and other things, like apple juice and pumpkin, to his formula to soften the stool. When that didn't help, we had to give him an enema. Fortunately his problem finally worked itself out just in the nick of time because as he got close to release age he was turning wild—bitey, not liking people, and not wanting to be caged. If he hadn't come around, he would have been euthanized for he would not have been happy in captivity.

Silver was another special one; she was born without a lower jaw. We didn't notice this until her teeth came in, and we began to offer her solid foods, at about four weeks of age. She came in at

Me with Silver

about a week old, sucking on a bottle wasn't a problem, chewing food would be impossible. Having had her this long, we were all very attached. There was no way we would euthanize this cute baby if we could find an alternative. We ground up her rodent chow in a coffee bean grinder, mixed that with baby food fruit, and found she could lap up this delicacy quite well. It wasn't easy to persuade management to save a squirrel with no jaw, but I argued that this would be a great way to teach kids about squirrels: she liked people, and could go to schools on education lectures. Plus with this genetic disability we were pretty sure there were others that cause her life to be very short. I bought her a nice big cage with my own money, and we kept her in our office.

We named this squirrel Silver for the unusual color of her fur. She became a wonderful part of all our lives—volunteers and paid staff alike. The break room was next to the office, so every day she had lunch with all of us. This was a special perk for volunteers, having lunch with a squirrel. I still have a piece of paper that I printed out when she played on the computer and typed a cryptic message. I also printed a copy of her belly while she played on the scanner. She was very much at home in the office with all of us.

Unfortunately, along with the genetic flaw with her jaw, when she was a year old, Silver also developed an inability to swallow, and we knew she probably would not live long after that. It was an extremely hard thing to do, to end her life peacefully with an injection ... but we knew we had given her a fun-filled life. And she had given much back to us—which we would have missed if we had ended her life when we had first discovered her jaw problem.

Chrissy was the last squirrel I raised before I left the center. When she came in at about three weeks of age, I noticed she couldn't use her left foot well; she limped a little as she moved around. My first thought wasn't, "Oh no, she won't be able to make it in the wild." Instead, it was, "Yippee, she won't be able to make it in the wild. I'll have a new Wheezer!" (Wheezer, who has his own chapter,

Ginny and Silver: One of the perks of being a volunteer, having lunch with a squirrel

was a disabled squirrel I'd kept as a pet; he'd died just a few months earlier.) I didn't express this thought to anyone else, for it is not the standard wildlife rehabilitation procedure to keep a disabled animal. Though it was becoming standard procedure for me, I still had to be discrete about it. At most rehabilitation centers, these animals were euthanized. I do think some animals are better off euthanized, for they have an innate will to live a wild life and are not happy in captivity. But I have had several disabled squirrels and know that, even though they don't make the best pets, keeping them can work if they choose to be in that situation.

Since I thought there was a chance of keeping this limping squirrel, I started thinking of names. I was a fan of the movie *Phantom of the Opera* and had just watched it for the third time. The phantom loved Christine as I loved the squirrel, so I named her Christine—Chrissy for short.

To my great (if selfish) disappointment, as Chrissie got older, her bad leg got better. I still wanted to keep her and probably could have gotten away with it (making some excuse for her inability to go free), but it just wasn't right to prevent her from being who she was if she could be a free animal. So I started her on a course of Squirrel Training 101. I had been caring for her at home because of her weak leg, but at this point I knew I would be leaving the center in October anyway, so I just kept her at home until the time came to release her.

Of all the squirrels I had raised and released until then, I'd never been able to release them on my property because I had several cats. I know they have to learn about predators, but several at once on the first day would have been overdoing it. I knew of other rehabbers who were able to release squirrels on their own property or somewhere close by, and I always envied them for it; in daily observations, they could watch the newly released animals adjust to the wild—and could also put out food and water as needed until they learned to find enough on their own. This is called a soft release, which is much better for the squirrel. But I'd never been able to do it that way. I'd read many books about squirrels, and until releasing Chrissy, I thought I had a good understanding of what it took to be a squirrel. But with Chrissy, I was finally able to watch a squirrel learn the post-release ropes, and it was quite an experience for me.

My opportunity to see Chrissy through the whole process came because earlier I'd lived in a cabin in the woods off the grid (no electricity) for two years. I asked the owner of the cabin for permission to come back, stay a while, and release my little squirrel there, and she said it would be okay. To begin, I set Chrissy up in a small walk-in closet with a window. It was a bit smaller than the spare room she'd occupied at my house, but it was larger than her cage. As I lay in my bed in the next room, I could hear her pitter-patter as she explored her new space. I gave her many "outside" objects that she'd be encountering once she was free—tree branches, boxes of dirt, moss, pinecones, bark pieces, and other woodsy things—and after five days, I opened up the window in her room so she could go out. Her window was four feet off the ground, and I put a tree branch from there to the deck floor beneath it. Timidly, she made her way down the branch. She stopped once as if to reconsider her route, looked back in the direction of her familiar home, and then continued down the branch to the deck. She scampered around, smelling things, for a few minutes until she found her way off the deck and onto the dirt forest floor. She wasn't there for more than a minute before she found her first tree and zipped thirty feet to the top. I felt all the pride of a mother seeing her kid accomplish her first "big girl" adventure. I wrote in my journal:

> *Day One:* Chrissy has been enjoying her new surroundings. What a beautiful array of new smells, fresh dirt, and more trees than she could climb than in one day. Her expressions and behavior told a story of wonderment and joy as she scampered up and down the fifty-foot trees. Occasionally my heart would skip as she would slip and

Chrissy learning her new wild skills

catch herself quickly from falling. Once she did misjudge the strength of a branch as she climbed out to its end. It gave way, and she fell like a parachute, all four legs spread out parallel to the ground. Plop, she landed on the soft earth and leaves. She seemed a little surprised but soon scampered up the next tree!

After an hour or so of climbing, Chrissy began to explore new foods. She dug things out of the ground and nibbled on moss and leaf tips. I was amazed that she seemed to know what to eat. As she continued to explore her new world that day, I wondered if she would ever come back to the safety and warmth of the cabin. There she had food and shelter if she wanted it. As she kept exploring and not returning, I did think she might have left the cabin for good. Then I saw her coming back; she climbed up the branch to her room, helped herself to food and water, and then she was off to explore the outdoors again. As night fell, I was relieved to know she could find her way "home" if she wanted to.

Day Two: Chrissy's behavior has become more serious. There was less play, and she seems to be in survival mode now. She spent more time hunting for food. She even buried some things. If she'd had a normal squirrel upbringing, she'd have been outside with her mom for the past two months since being weaned and would have learned how to find and store her food for the winter and build a nest. This is a very important part of survival, and she has a lot of catching up do; she's been disadvantaged. The most important part of survival, I believe, is nest building. Without shelter from cold and rain, the young squirrel would surely die.

Day Three: As I've watched her try to build a nest for the past two days, I'm thinking this might be a large part of why half of young squirrels don't live a full year. I watched her tirelessly run to the end of a branch, chew off a twig and carry it back to a larger branch, the next possible nest site. It was sad to watch as she would place a twig or a small piece of moss (that she had just dug off a tree) into her nest spot, only for it to come loose and float to the ground. Even sadder was to watch her young face stare at the piece as it fell, and she had this "oh shit" look in her eyes. I hung two well built, pre-made squirrel houses in two different trees, hoping she would choose one for a nest, just in case, but so far she has ignored them. My thought is that where an older, experienced squirrel might see the human- made house and know, "Great, I won't have to build a nest this year!" the young little squirrel brain only tells her to build a nest.

I have also noticed in the past three days Chrissy's natural alarm system has kicked in. She is a little less trusting of me now, though she did jump on my shoulder briefly today. She will go into fear mode at the slightest new sound or movement. In that state, she seems to turn to stone, just freezes, and listens intently. When danger has passed she is on the move again.

I will continue to supply food and water to Chrissy as she learns through the winter. It is so hard to watch her struggle with nest building and trying to find food. I suppose this feels much like a mother sending her child off to school for the first time. Or perhaps to college, where they have to fend for themselves for the first time! I can only hope that, with a watchful eye and loving care, she will beat the odds.

A year later, I can report that throughout that Oregon winter—daily and later twice-weekly—I made the trek in my car up and down the several miles of rocky, unpaved road to the cabin to leave food and water there for Chrissy and perhaps catch a glimpse of her. A sacrifice on my part? Not really—that was easy compared to what Chrissy must have gone through in her first winter on her own.

And it was worth it. A year after release, I spotted Chrissy in the area, living a normal squirrel life there!

I have to insert a note to my reader here: The next time you get upset at a squirrel raiding your bird feeder, put yourself in its paws. The life of a squirrel is a constant struggle for survival—especially for a juvenile. It's known that only 50 percent of baby squirrels make it past their first year. Why? Because it takes a young squirrel at least *six months* to learn the basics of squirrel survival. That's just the basics, and from then on it's a race to see if those basics can manifest as skillful survival techniques before starvation or a fatal injury occur.

So when you feel angry at a squirrel robbing your bird feeder, you might think of Chrissy and consider putting out some extra food. You may be helping one squirrel reach her first birthday!

Not Quite "Retired" Yet: Scrat, Samantha, and Bravo

Since I left the rehabilitation center, I've still managed to receive an orphaned squirrel to raise every year. They've kept my squirrel love alive; though it could never really die, having this opportunity to save them just adds intensity to the love I feel for these creatures.

My first year in "retirement," I got a call from a woman who was raising two babies that had fallen from their nest when the tree they were in had to be cut down. No one knew about the nest until it fell. I don't remember how she got my name, but I gave her some advice over the phone. About a week later the little female was injured during a rough play session with her brother, and we decided that it would be best to separate them to avoid further injury. I took Scrat, who was then about eight weeks old and just starting to eat solids. The woman kept the little female with the sprained leg. When Scrat was about twelve weeks old, I soft-released him back in the yard where he had been born. The family who had found them kept an eye on him after that, and I was able to enjoy Scrat's adventures with them. It was so much fun watching the woman, who found and raised him from a pinkie, see him climb trees for the first time. We both were taking photos and videos of his every move. Since his sister had the injury, she developed a little more slowly, but in two weeks she, too, was out and about with her brother. Last I heard, both were doing very well. With the help of their surrogate human parents, I am sure they made it to their first birthday.

Samantha came to me the next year, courtesy of my friend Kim, another rehabber. Kim had actually taken the baby squirrel in first but had her hands full at the time with baby raccoons. She knew how much I loved squirrels and asked me if I would like to take care of little Samantha. I am sure she already knew the answer! Samantha was unusual in that, when she was about eight weeks old, she weaned herself, and soon after that, she paced the window of her room, clearly longing to be free. As I've said, some squirrels "wild up" quickly, and some have to be pushed out the door. I was pretty sure that since Samantha was so feisty, she would make it fine in the wild. When she reached pre-release age, I took her to a large chain-link cage in the woods behind Kim's house that was perfect for a pre-release enclosure. Soon Samantha became very protective of her territory there; if I came too close to put in new food or water, she would run toward me, making aggressive warning sounds. A warning if not heeded would result with her teeth sinking into my skin.

I built a separate entryway so I could safely slide in fresh food and water and keep her from getting to me or escaping when I did that. Because some dogs and cats frequently came around Kim's

Samantha's first introduction to a pine cone

house, it wasn't the best place to let Samantha run free, so when the time came to release her, I trapped her in her night cage (a small animal carrier) and carried her to the next-door neighbor's property. The neighbor had squirrel-feeding stations, so we could make sure Samantha had food and water while she improved her survival skills.

On the day of her release, Samantha spent only a few minutes looking around after coming out of her carrier, and then she zipped straight up a very tall tree. I think it must have been seventy-five to a hundred feet tall. Near the top, she was just a dark dot—and, for me, the definition of "fearless." I am sure she adapted very well. Kim and her neighbor reported "Samantha Sightings" for several months, but after that, it was hard to tell her from the other squirrels.

And next year, I got a call late in the evening about a newly found squirrel. The finders had called the local rehabilitation center where I had last worked, but they were closed for the day. Then they looked on the Internet to find a local licensed rehabber, and my name came up. As soon as I hung up the phone, I drove out to pick up the squirrel. And as soon as I saw him, I was in love. I even already had a name picked out. I was reading a delightful book, *A Favor in Return* by Greg Lindenbach, about a human's intriguing adventure with a family of squirrels; the youngest one was named "Bravo." This tiny, three-week-old squirrel looked to me like a "Bravo." (I only wished he could have talked like the little Bravo in the book.)

And so I was off again on the labor-intensive adventure of raising a squirrel. During the first twenty-four hours, I fed him the hydrating liquid every three hours. Next, I slowly introduced the squirrel formula. At three weeks of age, he still hadn't seen me; it would be another week before the baby squirrel opened his eyes. That day is always very special; it's very heartwarming to know they are seeing their world for the first time and that you are the mom they see.

The task of raising baby squirrels is all-encompassing during the first eight weeks. If you start with a pinkie, you are on call 24-7 with feedings every two hours for a week, followed by feedings every three hours for the next four weeks. You get very little sleep during this time, but after eight weeks, you look back and wonder where the time went. They grow up so fast. Some are ready to be released as early as twelve weeks, but then there are others, late bloomers, who don't feel ready until they are six or eight months old. And when you have known them as babies, even that seems early. It is a hard life for squirrels, out there in nature, and you feel an ache in your heart, knowing they might not make it. You want them to be at least two years old before letting them go, but by then they would have become too used to a captive life and unlikely to adjust to the wild life. So you *have to* send your babies out there when they're still small.

Since I have raised many lone squirrels, I have become very experienced at teaching Squirrel Training 101. Lone ones, who don't have another squirrel to play with and see what squirrels are by having another to grow up with, are disadvantaged. But I'm happy to say that all my lone, released squirrels have been observed many months later, still fending for themselves, so I am pretty sure I taught them well.

Bravo was a lot of fun. Of course, during his babyhood, he was delightful. I could hardly wait to wake him for his bottle-feeding and have a chance to hold and cuddle him. Once his eyes opened, he gradually started to play. I gave him little "stuffies," small stuffed animals to jump on and wrestle with. All baby squirrels love these; for a lone squirrel, it is at least something like having a sibling to play with. Like other squirrel babies, Bravo also liked to do "hand-play," in which he jumped on my hands and played at attacking them.

By the time he was approaching eight weeks and weaning time, Bravo already had his own room and a large cage. I furnished his room with several trees, firmly anchored in Christmas tree holders, to give him the opportunity to learn how to climb and jump from tree to tree. You might think this skill would come naturally to a squirrel, but actually they have to learn climbing by trial and error. At first, he wasn't sure exactly how to get around on his trees, and as he explored, he tended to misjudge his weight, trusting a branch that couldn't hold him, or he'd leap and totally miss the branch he was aiming for. But he didn't have too many of these mishaps before he became an expert.

I also gave him some large litter boxes of dirt so he could practice burying nuts. That is always a lot of fun to watch. Squirrels search for the best spot, dig proficiently, place the nut in the hole, push dirt on top, and then—the cutest part—pat it down.

And then I include a course on how to drink water. "That needs to be taught?" you might wonder. Yes, there is a knack to drinking without getting water up the nose. You can hook onto the side of the cage one of the water bottles made for small animal cages, and I once did that, but then I realized it wouldn't teach them how to get water in the wild. Now I give them a small dish and add some pebbles or small rocks so that when they put their mouth in to drink, the pebbles prevent their nose from going underwater. *Voila!* They learn to drink.

While squirrels are going through the weaning process, another necessary task for mom is to offer them natural foods—leaves, moss, seeds, pine cones, acorns; anything from the outdoors that a squirrel would eat. I always thought it fascinating that, when offered something wild, a squirrel raised in captivity will always go right for it, even bypassing the favorite store-bought foods.

At nine weeks, Bravo was weaned from formula and eating solid food very well. There is a week or so during the weaning process where it seems they might starve; they don't appear to be eating anything but don't want formula anymore. Somehow they get through that, though, because they continue to grow and have exceptional energy.

Another thing that happens during weaning is that the babies begin to have aggressive moods. Even though a rehabber might

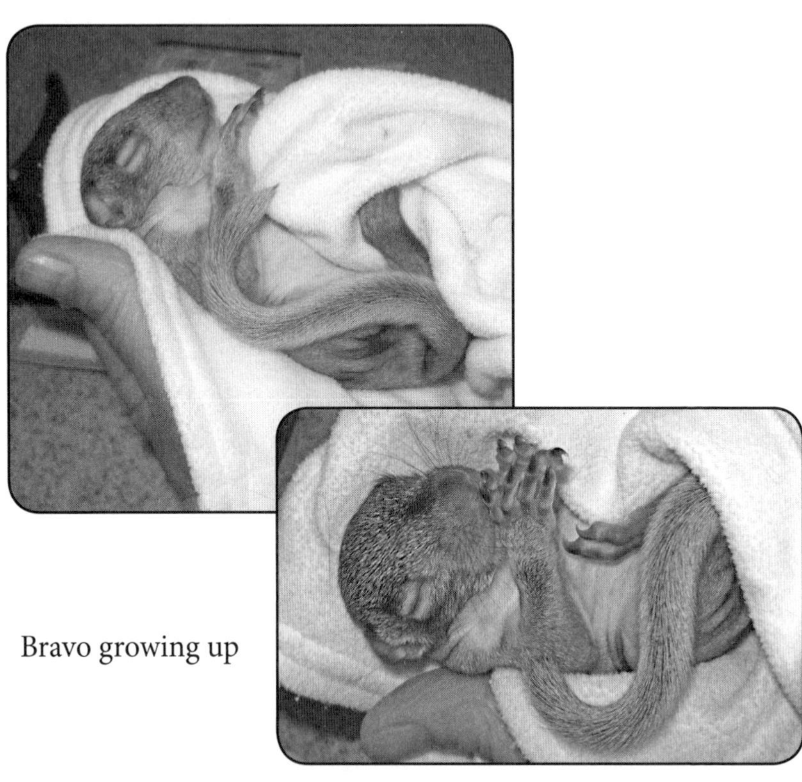

Bravo growing up

suffer some hurt feelings at this, it's really a good sign because it shows they are "wilding up." It usually happens just occasionally. Some young ones begin to sense the need to protect themselves and their food. One time I am with the little squirrels, and they are playful and carefree. Another time the natural wild behavior kicks in, and they bite me for intruding on their space. Then they turn into your baby again and want to play. These are all good signs that the babies' natural instincts are developing. Once the babies are on the road to freedom, the rehabbing experience takes on a mixture of happy and sad feelings, and then the hardest time comes, when you must let them go.

Bravo did very well. For his pre-release, I took him to the property of a friend who had worked with me at the rehabilitation center. Ginny had released squirrels many times and even had a large pre-release cage in her yard. In the days after I put Bravo into his pre-release cage, I went there several times to visit him—at first every other day and then just once a week. I did this hoping to make his separation from me easier.

On the day I released Bravo, my heart seemed to be breaking and flying at the same time. When I went into his cage, he jumped onto my shoulder. I opened the door and walked out with Bravo still on my shoulder. At first, he didn't move. He just looked around. Then he slowly climbed down my body to the ground. And then after just a few seconds, he suddenly took off and zoomed all over the yard, squirrel-style, so fast my eyes could barely keep up with him. Ginny and I sat in chairs in the yard for a couple of hours watching him. She had about ten other squirrels that came into her yard daily to scavenge the seeds that fell from her bird feeders. I wondered if poor Bravo would be attacked when he first encountered one of them. It was nearly an hour before one of the other squirrels appeared, and when that happened, Bravo began creeping up on it. I worried that he might be about to have his first bad experience in his new world and squinted my eyes, wanting to close them when he reached the squirrel. But it all went better than I thought; Bravo chased his first squirrel, demonstrating that I hadn't raised a wimp. I worried no more about his ability to take care of himself.

I hated to leave Bravo on his own but felt assured that he was adjusting well, and it was a comfort to know that Ginny was there. She put out a dish of food for him every day for several months to make sure he got enough nourishment while he learned his way around his new neighborhood. She told me she didn't think he even went back into his cage at night; from day one, he was ready to be a big boy and sleep in the trees.

As I write this, it has been a year since that release, and even though it has become more difficult to distinguish Bravo from all the other squirrels, Ginny is pretty sure she still sees him from time to time.

Chapter 8

WHEEZER: My Soul Mate

Of all the squirrels that have scampered into my life, one put his paw print indelibly on my heart.

Meet Wheezer: He came into our center in very bad shape. He was about six weeks old, fully furred, and his eyes were open—well, just barely because he was too weak to open them very far. He had fallen from his nest and been found by hikers in the woods. By the time they brought him to the center, they had named him Wheezer because he was wheezing as he breathed. This was partly due to an upper respiratory infection and partly—as I later found out—because in his fall he had knocked his nose crooked.

Despite our policy not to name animals that will be released, occasionally one comes in with such a personality that it names itself—and in Wheezer's case, he came in with a name and a personality!

Wheezer was so weak at first that I had to force-feed him milk formula, drop by drop, using a small syringe with a tiny rubber squirrel-sized nipple on the end. Each feeding time, I was afraid that when I opened his box I would find him dead. If he had curled up like most baby squirrels do when they sleep, he would have been the size of a tennis ball. But he was too weak to curl. Instead, he lay on his side in a half curl. I cuddled small stuffed animals around him, hoping to give him a sense of security. Then one morning when I opened the box, he jumped out and walked around! I was so thrilled that I called Chris, my boss at work, and said, "Wheezer's up, Wheezer's up!" I was over the top with excitement.

He still had some weak days, but he did become a healthy squirrel baby. In a few more weeks, he was weaned and eating walnuts just like a regular squirrel. But soon I noticed he wasn't a regular squirrel. His teeth were growing out of alignment, which meant the top two could continue to grow and curl around and into the roof of his mouth while the bottom two would grow up over his top jaw and into his nose. The only way he could survive would be for humans to trim the teeth every four to five weeks. I knew then that Wheezer could never be released.

Earlier, as you may recall, I had persuaded Chris to save Peanut from being euthanized (Wheezer came in about a year after Peanut died). I wasn't sure if I could do it again with another squirrel, but Wheezer and I had really bonded during his early stage, and I knew I had to present my case with conviction. I'm not sure what I said, but it was enough to save Wheezer. Actually, I think Chris just knew me well enough by now to know that telling me no was not an option.

And so Wheezer came home with me. I built him a two-story squirrel condo to stay in when I wasn't at home. When I was home, he was out and about with me and my five cats. He fit right in. Because the cats had known him since he was a baby, and vice versa, they all thought we were a natural family.

In the mornings, I would let him out of his cage, and he would follow me around as I did chores, or he'd ride on my shoulders until I fixed his breakfast. He'd busy himself with eating and then with burying nuts around the house. I kept finding nuts in the weirdest places—folded into clothing I had left on the floor, behind couch cushions, or in my shoes (that made an uncomfortable fit).

Wheezer when he came in and was sick

Wheezer the day he jumped up after being sick

When I left for work, I would pick Wheezer up and carry him back to his cage to stay while I was gone. This was for his safety. I wasn't totally sure that all my cats could be trusted. But while I was at home, he could stay out as long as he wanted. He usually would run around for a couple of hours and then put himself to bed for a while. He did like his condo.

The house where we were living was being sold so I had to move. About the only place I could find that would take five cats, a squirrel, was small cabin deep in the woods, the one I mentioned in an earlier story where I later kept Chrissy for her release. Wheezer was the last on my list of animals to move to the cabin; I wanted to wait until everything else (including the cats) was at the new place before I took him and his condo over there. Then, as it turned out, I had added on so much to his cage that it wouldn't fit through the door! It was too complicated to tear apart and put back together so I had to do a rush job building a new one at the cabin. By this time there were only two days left before I needed to be out of the old place; I'd already been given a week's extension. To add to the problem, because the cabin was off the grid, I couldn't use any power tools or turn on the lights at night to see what I was doing as I built Wheezer's new condo.

I decided to make a PVC plastic pipe cage for him. That way, all I needed was a small saw, pipes, and glue. Once the frame was built, I could use plastic cable ties to attach hardware cloth (a very sturdy wire mesh) to the frame, to keep him in. For those two days, I'd go to the cabin after work to assemble Wheezer's new cage and feed the cats, and then go back to my stark, empty house and Wheezer. By the time I finished work at the center and arrived at the cabin, I only had a couple hours of daylight left, and this limited my assembly time. I remember squeezing in a few more minutes by using a kerosene lamp as I strained to see what the heck I was doing. Good thing I wasn't using a hammer and nail—not sure my thumbs would have survived! I managed to finish the project

Wheezer greeting my cat, Shadow

in two days and with no time to spare before I had to vacate the old premises.

Not surprisingly, the new cage was far from perfect. It looked as if it had been built in the dark! But the door closed, and Wheezer could be safely kept inside when he needed to be. Since his new condo wasn't very square, the hooks I used to latch the door shut did need some extra reinforcement. I solved this by tying a shoestring around the door and frame to squeeze tight some possible escape holes.

We finished the move and settled into the cabin. Wheezer loved it there. When out and about, he had two lofts to climb in and wooden walls he could climb up to where he could hang out on windowsills and look outside. I was a little concerned about the wood-stove, though. It was winter, and the stove heated the cabin. The exposed pipe was very hot most of the time, and it was a squirrel's leap from one of the windowsills. I thought Wheezer might think it was a tree and jump onto it. To block him from going from the windowsill that was far from the stove-pipe to the next one that was closer, I placed a cardboard barrier between them, but Wheezer soon managed to do what squirrels do—he climbed up the wooden wall and around the barrier. Still, I guess he was smarter than I thought because he never attempted to jump to the pipe in the year and a half that we lived there. Maybe, being smarter than I was, he just knew that trees aren't supposed to be hot.

One problem I did have with the cabin was at bedtime. The lofts were like squirrel mansions. Each was a nice climb up to about three feet from the ceiling, and provided soft blankets and cubbyholes to snuggle in. For him it was like being high up in a comfortable tree. So there were many nights when I had to find him in one of the lofts before I could put him to bed in his cage. I didn't like having to worry about him being out loose while I slept—and when maybe not all the cats were sleeping. On top of this, there was the possibility that he would find the cat door and escape to the outdoors.

And so before going to bed, I had to climb up to each loft in search of Wheezer (well, only one of them, if I was lucky enough to find him in the first place I looked). In the loft, which didn't have much headroom, I'd have to crawl around searching for a lump in a blanket. I would gently tap each lump. If it moved, I'd found him.

Most of the time Wheezer seemed to realize when he was found that he had to go to his cage, and he let me pick him up with one hand and crawl to the ladder. Resigned to his fate, he held on to me as I climbed down and put him in his cage nest. It probably wasn't his favorite routine, but I felt a special closeness to him—kind of like putting my kid to bed after he fell asleep on the couch at bedtime.

One night, I came home from work and found that Wheezer had somehow pushed his door open. He was gone. My first panicked thought was that he'd gone out the cat door and was lost in the woods. I ran outside calling his name. It was very dark, and I couldn't see anything, but I was sure he would come to my voice. No Wheezer. I went back inside to survey. My next horrified thought was that the cats had decided he was prey and got him. I searched the place with my flashlight looking for a limp, still squirrel. Nothing.

Then I thought—oh, of course, he was in one of the lofts! I searched them both, thumping lumps and searching under the blankets. No Wheezer. I was about to give up the search until morning and sat on my bed trying to calm down. Then I heard a slight stirring noise behind me in Wheezer's cage. I looked up and there he was, popping his head out from under his blanket! He had put himself to bed. I think I became superhuman then; my emotional state went from the depths of despair to sky high joy faster than a speeding bullet.

After a year and a half, we all moved to a nice house with hot and cold running water, lights that came on at the flip of a switch, and an indoor bathroom. I felt like a millionaire. I had loved cabin life and learned so much—like appreciating the miracle of electricity. I owned very little. Furniture had never been a top priority with me. The first thing I bought was a nice three-hundred-dollar cage for Wheezer. It had to be assembled, but with the help of a friend, it was up in a day. Wheezer loved it. It was about twice as big as the plastic pipe cage he had just left. I guess he got a new fancy home, too, and seemed to know it.

We all loved settling into such comfort. Wheezer spent a lot of time exploring all the wonderful climbing and hiding places in this new seventeen-hundred-square-foot domain. I think his favorite part was the many rooms of lush carpet where he could bury nuts. He spent a good part of his morning digging "holes" in corners of the rooms or behind doors to hide his nuts.

One day after four months of settling in, I went into his room to let him out of his cage and found him at the door waiting but saw lots of blood on his blankets. I quickly opened the door to get him but he scampered out before I could catch him. I found him in the kitchen and picked him up to look him over. He was pretty squirmy so I couldn't find where the blood had come from, but I knew it must be something serious because there was a lot of it. I grabbed a cat carrier and put him in and took off to the center where he could be sedated enough to check him out.

Once he was sedated, I found he had cut open his testicle somehow, and some of the insides were coming out. I rushed him to the vet and luckily they took him right into surgery. I held him as the vet tech put a tiny cone connected to the sedation gas over his droopy head. The vet did an expert job of placing things back in and sewing him up. As he was on the last stitch, I noticed Wheezer wasn't breathing well. The vet tech tapped on his chest a little, but it was too late. Wheezer had stopped breathing. I was in shock as I took him off the table and gently put him back in his carrier. I don't remember going home, but I do remember I couldn't stop crying.

For days, all I did was make memorials to Wheezer. I made a slide show, wrote a book, made a framed sign with his pictures saying "Wheezer's Room," and hung it in there. As time has gone by, I know now more than ever: Wheezer was my soul mate. Can humans have animals as soul mates? You bet they can.

Wheezer, my soul mate

Don't feel too sad about Wheezer and me. I still get sad missing him, but I also feel so blessed by the time he spent with me in his short three-year journey on this planet—and I have no doubt I will see him again!

Chapter 9

Lenny the Lynx: My Best Bud

Lenny was a "small bus" Lynx. This was a term of endearment that Chris gave to our disabled animals because, usually, small yellow school buses were used for kids needing special care.

Lenny was born on a lynx breeding farm in Minnesota. It is difficult for me to understand why those places are allowed to exist. In Oregon, it is illegal for me to keep a disabled squirrel as a pet because it is a native animal, yet almost anyone can get a permit to buy and keep exotics like lynxes and lions. It makes no sense to me.

Lenny's previous owners had bought him right after he was weaned and then kept him only a week before they decided he was too hard to handle. Then he was sent to us. Because he had been bred in captivity, legally he could not be released back to his native territory—even though after meeting him, I had no doubt he could have been trained to return to the wild. He was quite the spitfire.

The night the baby lynx was expected, I stayed at the clinic, awaiting his arrival. He was brought in around ten in the evening in the animal crate he had been shipped in. I looked inside to see a very frightened, fluffy, yellow ball of fur, hissing and shaking as his wide eyes stared at me from as far back in the crate as he could get.

He could have fit into the cup of my hands (if he had held still, and if my hands had been gloved). My heart went out to him. He had been taken from his real mom as soon as he was weaned and put into a very foreign situation with humans. He had bit and scratched his new captors and, I am sure, tried to run in search of his mother. I assume those humans had wanted a cuddly little exotic kitten and, being very disappointed in this little lynx's disposition, just got rid of him. I could tell he had been hit with hands, probably for biting, because when he crawled toward the front of the crate and I tried to pet him, he cowered.

I let him stay in his shipping home with his security blankets, hoping he would slowly adjust to his new location, I stayed with him overnight to try to instill in him some trust in humans. First I lay down next to the open door, and after about thirty minutes, he was ok when I put my arm inside. After a few hours more, he finally fell asleep next to my arm for comfort. I thought I had made great progress, and the next day he would already be on the road to being a friendly little kitten we could all bond with—but that was far from the truth. He had been so traumatized that his road to recovery would be a long one.

Every day, we all took turns being with Lenny. We kept him in a larger kennel inside the clinic so that he could have a secure home, yet come out and be with his human caregivers whenever he felt up to it. He did begin to play with toys after a few days but didn't like to be touched or picked up.

After a few weeks, he was put outside in a big enclosure and started in a training program. Everyone thought because he was so young he could be easily trained for education programs. But months went by, and he made little progress in tolerating being handled. We did manage to get a collar on him and then a leash, but he wasn't fond of that. Leash trained just wasn't something Lenny was going to be, at least not anytime soon. Finally, one of the trainers found that he could pick him up if he scruffed his neck, and once he was off the ground in this manner, Lenny would become very still. I

Baby Lenny

think it was an instinctual behavior because lynx mothers scruff their babies as they move them from place to place. I was involved with the training process when I had time, but mostly I would go in with Lenny after work just to be close and gain his trust.

After almost six months, the main trainer quit the program. The next day I came in and, ignoring the formal procedures such as clicking him with commands, I simply picked Lenny up and sat down with him on my lap. At first he tried to run, but as I talked softly and petted him gently, he gave in. It was as if he were a young child starved for love but not wanting to show it. He wanted a gentle touch, but his fear of hands wouldn't let us break through his tough exterior, and no one had been willing to challenge that. After this breakthrough, every day I would spend time with Lenny on my lap; he mellowed quickly, and soon I was able to take him out for walks on a leash.

Sheila, another trainer, had been successful with our Eurasian lynx and began working on Lenny's training skills, using several commands. All went well for about a year, but then he suddenly developed some neurological problems. He began to stumble and act as if he couldn't see; his eyes would sometimes dilate and not return to normal. An eye doctor examined him, but couldn't come to any conclusion about the cause of the problem. We took Lenny to a hospital to have an MRI done, but still there was no conclusive diagnosis. We had to assume that perhaps the circumstances of Lenny's early life—inbreeding or a poor diet—might have contributed to his condition.

And so Lenny became the small bus lynx. He was taken out of the training program, and I was glad about that, though not happy about his condition. Now he no longer had to "perform" for his meals or be told what to do and how to do it; he could just do and be who he was. Training isn't necessarily a bad thing—it is actually fun for many animals—but Lenny wasn't one who enjoyed it.

He became my best friend at the center. I spent every spare moment with him that I could and always put him to bed at night.

Best Buds—Lenny and Me

He had very little in his life except me. He was in a cage alone and not fond of his neighbor, a bobcat name Shoshoni. Shoshoni and I never bonded, but I always went in her cage for a quick game of chase with her on my way to Lenny. She didn't care much for pets or cuddles but did seem to enjoy me to "play-chase" her around her cage. Then in Lenny's cage, I climbed up into his house that was 4 feet off the ground, and he came and cuddled next to me. It was our special one-to-one time that we both cherished.

When I left the rehab center after being there five years, I felt there needed to be a gradual transition for me and the animals I had special bonds with, especially Lenny. Of course to most humans, including those in charge of the center, that would seem impractical and unnecessary. Why would an animal need a transition? But I think Lenny did.

I was caring for a disabled squirrel at home, so I went to the clinic at least once a month to get supplies and trim her teeth. I would also visit Lenny. He was always happy to see me. Everyone at the center was too busy to let me into his cage, so I'd just try to pet him through the chain link. He'd rub against the fence and purr and purr. When I left, he would run after me, up to the barrier of his fence and then cry and cry. It broke my heart.

The one good thing was that Caryn, the woman who took my place, loved Lenny too, and they bonded. It felt good to know that someone still cared about him and gave him love.

After several months of visits, I began thinking that maybe I was doing Lenny more harm than good; I hated to see him cry when I left, and I knew he was getting attention from Caryn by then. So I felt that maybe it was time to leave him for good.

About two months later, I heard that Lenny was not doing well. His eyesight was getting much worse, and he was definitely going downhill for some reason. Then I heard that he was going in circles when he was on the ground, and even fell from his house because he couldn't see and was disoriented. A wider ramp was

Me after taking Lenny for a walk

built, from his house to the ground, in the hope that this would prevent another fall.

I felt the need to see Lenny again, and on my next visit, I went out to his cage. He was sleeping in his house, which I was told he was doing more and more. I called his name repeatedly with no response. Finally he lifted his head slightly, mewed twice, and then fell back to sleep. I don't know, but I think that somewhere in his memory he recognized my voice and said hi.

Not long after that, the staff made the decision to euthanize Lenny. He had no quality of life anymore, and I don't argue with the decision, but I am so glad I wasn't still working there and didn't have to be the one to make the decision to end his life. Poor Caryn, I know it was terribly hard for her.

While this might seem egotistical, I sometimes wonder if my not being in Lenny's life, not putting him to bed every night, might on some level have worsened his condition. I wish I had been allowed to come every night to put him to bed for several months after I quit, then slowly withdraw my presence as Caryn took over. It doesn't matter now. I know my Lenny is flying high somewhere with my dear little soul mate, Wheezer the squirrel, and that thought does make me smile.

Chapter 10

Baby Bear: Best Release

My fifth and last year at the center started with a spectacular experience—one that, in hindsight, I feel was a reward equal to all the heart and hard work I gave to the center.

On the first of January that year, I was at the center with a volunteer, Karen, when we got a call that an abandoned baby black bear had been found. It was about four hours from us, so the wildlife authorities in that area arranged for a state trooper to meet us at a halfway point with the bear. It was a cold and snowy day as we rode the winding roads, Karen at the wheel so that I could care for the bear on the way back.

As we arrived at our destination, it was growing dark, and I felt as if we were on a secret mission when we turned into the parking lot of a motel and pulled up next to the trooper's car. He handed me a small box; I thanked him and got back into Karen's car. We could hardly believe what we saw when I opened the box. We pulled back the towel and there was a small ball of fur no bigger than a softball. The little bear's eyes weren't opened yet, and his tiny ears were flat against his head. Obviously he was very young, but as I didn't know much about baby bears that young, I couldn't even guess his age.

Karen drove out of the parking lot, and I had the little bear in the box on my lap. We weren't on the road long when the cub began to cry. At first it was a soft whining sound, but soon it grew into a loud, high-pitched screaming "*rraaah*."

I put my thumb in the tiny cub's soft, toothless mouth, and he began sucking quietly. During the two-hour drive back, I did all I could think of to give him comfort, knowing the little one must really be missing his mother. I was sure he felt fear of being in a strange place without his mom and that he must be hungry as well. When a thumb in his mouth no longer satisfied him and he began to cry again, I picked him up and cuddled him in my arms. Rubbing his back seemed to calm him, too. By the time we arrived back at the clinic around nine o'clock that night, Karen and I were emotionally exhausted, and I'm sure the bear was, too. Karen offered to stay and help, but I thanked her and said she should go on home. I would be fine. I was just going to look up some information on baby bears and then would take him home with me so I could keep an eye on him.

I was extremely nervous about taking on the responsibility of raising this tiny precious being. All the other orphaned bears I had cared for in the past had been at least two months old when my care began, and I hadn't had much time to learn about raising such a young bear before we picked this one up.

I turned on the heat in the clinic's incubator, and once it had reached about 98 degrees, I put the bear into it, nested in a blanket. He seemed to be exhausted and went to sleep. Meanwhile, I filled a baby bottle with several ounces of a hydrating liquid. All baby animals arriving at the clinic were offered this electrolyte drink, first to hydrate and then to clear their stomachs to ready them for something new.

When the liquid in the bottle was warm, I began a familiar struggle—trying to feed a baby who was not pleased when an unfamiliar nipple was poked into his mouth. Most orphaned animals resist a nipple different from their natural mother's. With this

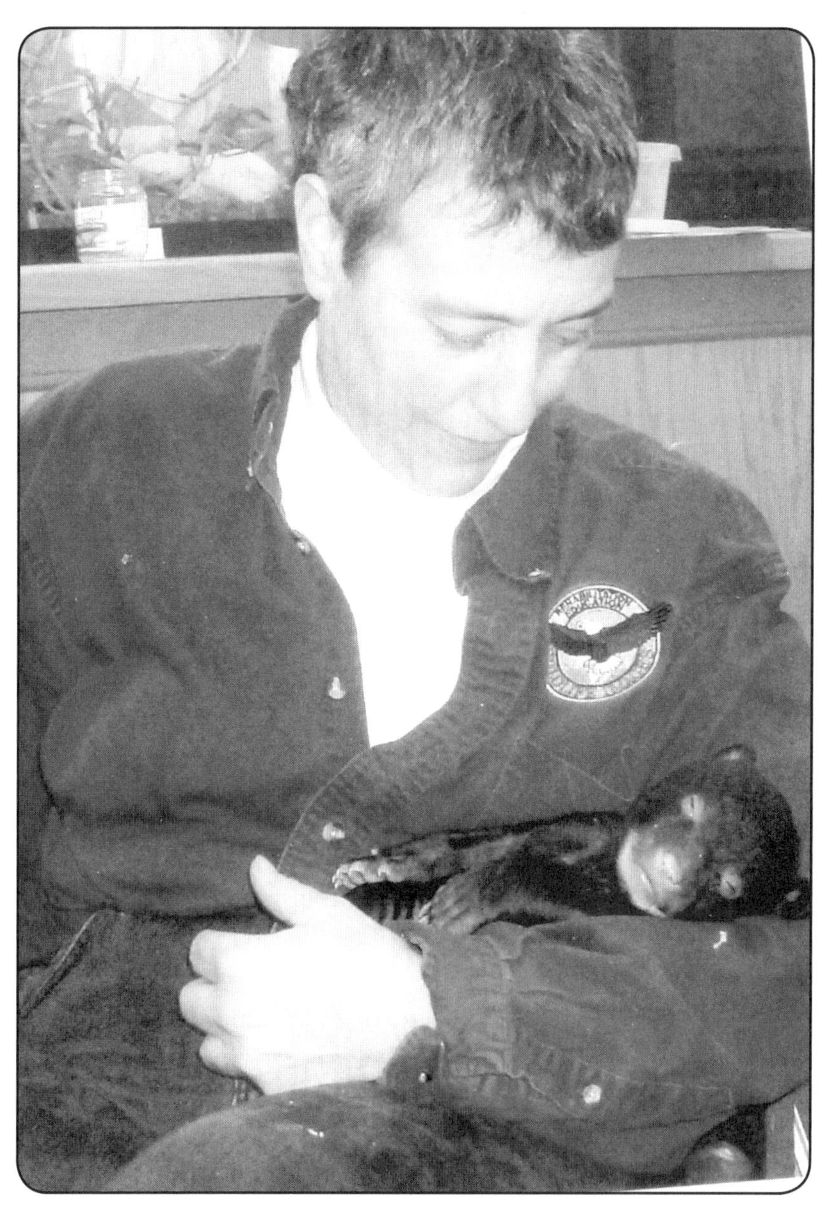

Baby bear soon after his arrival at our center

bear, I used a Playtex nurser with a plastic bag inside a hard plastic bottle. I left the cap off the bottom so I could press on the bag to squeeze the liquid slowly through the hole in the nipple to let the baby know that something soothing could come out of the strange object in his mouth. Between cries and struggles, he managed to swallow about two ounces of nourishment, but he took forty-five minutes to do it. That was enough. I rocked him to sleep and put him back into the incubator. As he slept contentedly, I looked up the care and feeding of baby bears in one of our books (*Idaho Black Bear Rehabilitation* by Sally Maughn, an expert in baby bear rehabilitation) in the clinic office, noting what seemed to be most important—formulas, measurements, and correct weights—and found enough information to help me feel more confident as I put the baby into his carrier and headed home for the night.

Once at home, I set up the bear's little carrier next to my bed so I could be near him during the night, ready for electrolyte feedings every two hours through the night. He contently snuggled in his soft blankets kept warm by a heating pad underneath. In the morning, I gradually introduced him to the baby bear formula mentioned in the book. This consisted of powdered Esbilac, a commercial formula for dogs, mixed with Multi-milk, a formula additive that increases the fat content (bear's milk is higher in fat than dog's milk). Again, it was a struggle to get the cub to suck on the nipple. I experimented with several different types and hole sizes, and finally, after two days, the bear began to look forward to his nipple and nourishment.

Our first nights together seemed almost unreal—well, at least for me. I felt incredibly special to be holding and feeding a real infant bear. It was very similar to caring for a week-old puppy, except that his paws were equipped with larger pads and longer claws. As I fed him the bottle, he moved his front paws in a swimming motion, as if to pad at the mothers breast. I tried to position my arm so he would press on it instead of "swimming" in mid-air.

The bear adjusted to his new den, the inside of a plastic pet carrier. Inside his blankets, he cuddled contently with his surrogate mom, a stuffed toy gorilla. As the days went by, he only fussed when he was hungry and when I had to perform bathroom duties. When they are very young, before their eyes open, many baby animals need help from their mothers in eliminating their waste; the mothers provide this help by wiping with their tongues. I used a soft damp cloth, doing my best to make it feel the same to the baby. I am a good substitute mom but not a perfect match.

I found Maughn's book invaluable for dealing with the baby bear. Several times when I had questions, I even called her, and she was extremely helpful. Using information in her manual, I concluded that this little cub must be only two to three weeks old. I wondered what could have caused the little guy to be separated from his mother because mother bears are fierce protectors of their babies. I think the only way she would have let her cub go was to have been shot, or perhaps when she was chased by someone, the baby fell unexpectedly and she wasn't able to go back.

That first night, I lay awake thinking what this little one needed and worrying about how to do the best I could for him. More than anything, I wished we could have had another baby his age so he wouldn't have to be raised alone; I knew his chances of survival in the wild would increase if he could be raised with a bear sibling. A brother or sister bear could teach him more natural bear behaviors—or at least baby bear behaviors that could lead to a wider range of adult behaviors. Even that wouldn't be as good for him as having a mother to teach him, but every little bit could help. Yet I also knew the chances were next to impossible that we'd get another orphan cub; in the five years I was at the center, we had only four orphans, and they were much older than this little cub.

The next day, a miracle happened: someone brought another orphaned cub to our center. This one had been found not far from our clinic, in a mountain forest. He was a black bear, but cinnamon colored, and looked about the same age as our little one. Because

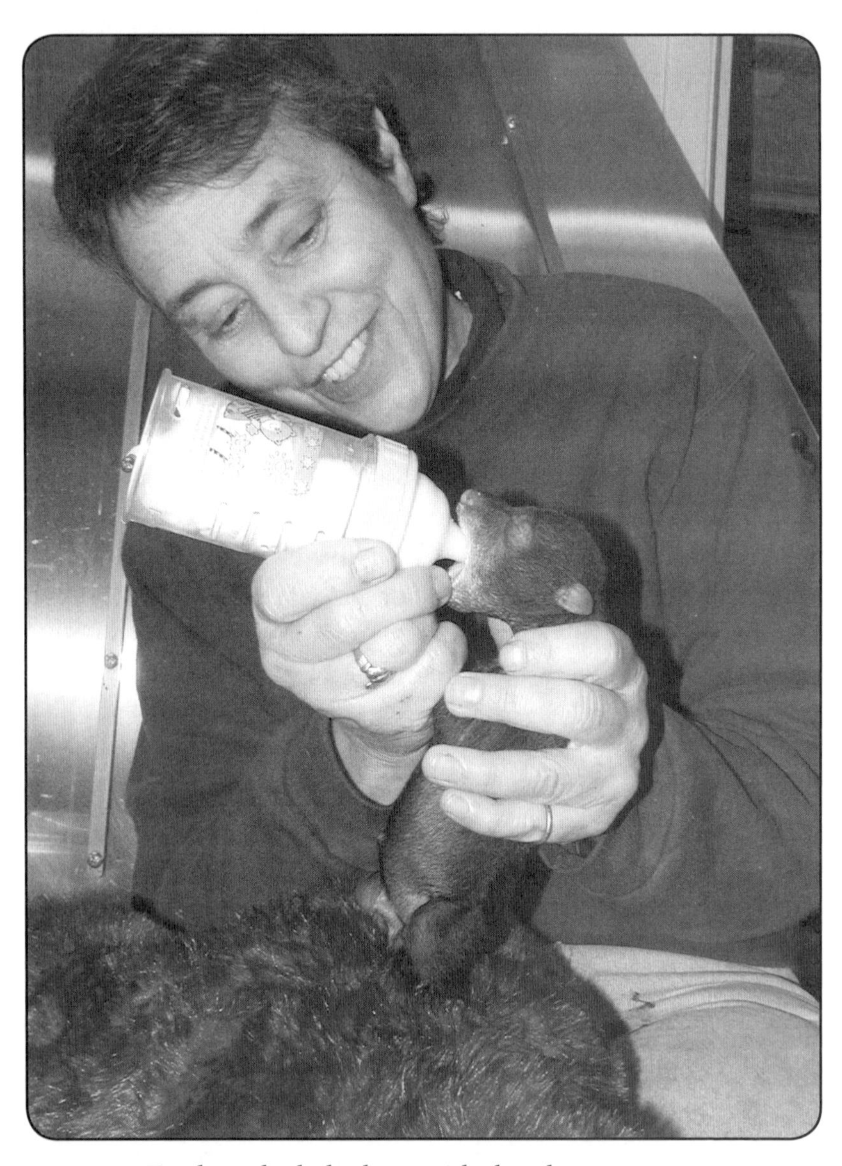

Feeding the baby bear with the playtex nurser

his paws were frostbitten, we knew he had been abandoned for some time. We warmed him up in the incubator and then put the two together. Immediately, they cuddled and fell asleep.

I called the vet to see what I should do for the new cub's frostbitten paws. He said, "Put some salve on them and hope the blood returns in the next day or two." But the next day, they looked the same. Luckily, the little cub didn't seem to be in pain, and he played with the other cub, which was very cute. They were too young to do much, but they did paw each other and cuddle. On Day Two, the frostbitten paws were beginning to smell, and that was not a good sign.

The director thought the second cub should be euthanized because, if they didn't heal, the affected paws would rot off. I didn't want to make that decision on my own, so I took the cub to the vet for his opinion. Our regular vet was out of town so I had to trust one I didn't know with this important decision; but she was kind and very gentle with the little cinnamon cub. He didn't seem to be in any pain, and apart from the problem with his paws, he was normal. I loved him and wanted the vet to say she could save the paws—not only for me, but also for the welfare of the first cub. The vet examined the paws closely and then gently, with tears in her eyes, said no, she couldn't save the paws. They were just too badly damaged. I understood the cub couldn't be released without paws, nor would it be fair to keep him in captivity without them, and resigned myself to losing him. I asked the vet if she could administer the injection because I didn't want to take him back to the clinic and do it, and she agreed.

As I returned to the clinic, I wondered if our little black cub would again feel abandoned, having lost his new sibling so soon. In my mind, I had secretly started thinking of him as "Trooper," the name our receptionist came up with after hearing that a state trooper had brought him to us. I say secretly because of the "no names" policy for releasable animals.

For the next two week, I lived on adrenalin and little sleep. My nighttime consisted of feeding the cub every two to three hours until one in the morning, followed by restless, worried sleep until the morning feeding around seven. Like any new human mother, I was ever attentive to any small movements or sounds little Trooper might make during the night. Then during the day at work, in addition to keeping an eye on Trooper, I had the responsibility of emergency care as injured animals came into the center and overseeing the care of our over one hundred permanent animals while directing the duties of our great volunteer staff.

After I'd been a bear mom for a couple of weeks and Trooper was about a month old, the director told me the cub would now be given to a real mother bear. He explained that wildlife officials in the area where the cub had been found were doing studies on several mothers and cubs. They put radio collars on the moms and then tracked their winter activity. My first internal reaction was, "Are you crazy?" for I thought that he might not be accepted by another mother. Of course, I didn't express this fear in front of the director, but I calmly responded, "Oh, that sounds interesting; has it ever been done before?" He said yes, but for me, that didn't answer the real question: Was it successful? Later that day, I spent hours on the Internet, searching to see if this had been done and how successful it was. I loved this little bear, and I wasn't about to have him killed by an irate bear that didn't want to adopt him. But what I found was amazing: Not only had it been done many times, but I didn't find a single case that wasn't a success. I even read of one mother that was raising two of her own cubs and adopted a third! It seems there aren't many other animals on the planet with mother instincts as strong as a bear's.

I went to bed that night with such a wonderful feeling that Trooper would have the best of all releases and be able to go back to a real mom.

Because the wildlife officials wanted to make sure the cub was in top condition before giving him to the adult bear, I was asked to care for him another week to make sure he was healthy and strong. And so, confident of the best possible future for the little cub, it was my sheer pleasure to keep him a while longer. During the next week, Trooper's ears started to pop up, and his eyes began to open. Surprise—a human mom! Gradually, he was becoming a little more active. He liked to have his belly rubbed, played with my hands, and kicked his paws and feet. He got better at crawling, too. One of the most noticeable changes was that he was no longer content to go to sleep with his stuffed gorilla and blankets; maybe now that he could see and hear, he became newly aware that something was missing. After our feeding and cuddle time he wasn't as content to be put back into the carrier. Now he would begin to scream after he realized he was alone. I couldn't stand this and took him out of the carrier and into bed with me. Once he was sound asleep next to me, I could gently put him into the carrier without waking him. There were times he did wake up and scream—and back to my bed he went.

It was during this week of transition that we got the word that the officials had decided it was time for the bigger transition to a real mom. They scheduled a release date on the next weekend. And now there was another happy/sad time to go through.

At six in the morning of the bear's release date, the director, Trooper, and I met the clinic supervisor in our center's parking lot to start our two-hour trip to take Trooper to his adoptive bear mom. I packed extra blankets and formula, as well as a few snacks for the humans. Most of the way, Trooper slept in my lap. I fed him just before we left, and once more just before our arrival. I didn't give him the full amount, because we wanted him to be a little hungry when he was put with the mother so that he would search for her nipple. After we got to the ranger station, they got into their car and led us for another half hour.

We parked as close as we could get to the den, and then the rangers got out and packed up telemetry equipment—a battery backpack and a hand held aerial—for tracking the mother bear. The plan was to find her asleep in her den and poke her with a long pole that had a syringe containing a sedation drug on its end. Once they were sure she was completely sedated, they would come to get us and our bear cub. They were gone about forty-five minutes. When they came back to our car, they gave us the disappointing news that the mother had just left her den so they couldn't sedate her. That meant we had to go back to the clinic and try another day, which would be the next weekend. I was a little disappointed because I had prepared my heart for the separation. But I was happy to have Trooper another week.

For the next week, the little guy continued to fuss at bedtime, but I didn't mind cuddling with him. During the day when he fussed after each feeding, I would play tickle games with him for a while and rub his belly to ease him into sleepy time in his carrier.

I was happy just to have another week to play with a baby bear—something I was sure would never happen again.

When the next weekend arrived and it was time to try finding the mom again, the routine was much the same as before. Again, we all waited in the parking area while the rangers could find and sedate the bear. But this time on their return to us, they came with a sense of urgency in their faces. When they got to our car, they said we had to follow them as fast as we could so we could get there before the mother woke up. I wrapped Trooper up in a blanket and followed everyone off to the wooded mountain. This was a hike like no other I had ever done: no beaten path, only knee high bushes and thin trees to climb between, over, and around. I was at a disadvantage, having to hold Trooper in one arm and use the other to hold onto trees for balance. After a half hour of this—all uphill—we reached the den, we were all huffing and puffing. Trooper was quiet for most of the climb, but during the last ten minutes, when the trek got rougher, he started crying.

Once we stopped and my tree-grabbing hand was free again, I put my thumb in his mouth and that calmed him down for a while. The den opening was very small. It seemed amazing what

Wildlife official putting the baby in den with new mother

the mother had done: she had dug out a large area inside a dirt hill, just big enough for her to snuggle in and then filled dirt back in at the opening so that it was just a bit larger than her head. All we could see was her head as she snoozed near the opening. The ranger got down on her belly and asked me for Trooper. He was crying as I handed him to her. Holding my baby, the ranger burrowed into the den up to her waist, and in about three minutes it was all over. Not only was Trooper with the new mom and the ranger back on her feet, but also to my amazement, Trooper was contently silent. He KNEW he was with a real mom. I felt like one big goose bump with tears in my eyes. The best possible release had just happened.

The rangers said they wouldn't know much more until spring. And even then, the interns watching the bears might not know which was which. Their study was simply to see if bears traveled much during the winter. I told them that was okay with me, that I was leaving the site knowing that "All is perfect, and I choose to believe it will always be a happy ending."

Afterword

Looking back on my life's journey, I know I will never lose the extreme gratitude and joy I feel for having been so totally trusted and loved by species other than my own. Of course it's possible to experience this with pets—animals who have evolved or been bred to be companions to humans—yet being accepted by a *wild* animal is truly a rare, humbling, and mind-altering experience. I wish for every human to feel that connection, too, at least once in his or her life.

I believe that in the Earth's ancient past, all creatures were connected to one another in this way. Somehow humans lost much of their knowledge and awareness of a reality that underlies our entire existence on the planet: oneness with all life. When we separated, other species also, perhaps, lost that sense. But as my life has shown me, trust and connection can be restored.

Keeping wild animals as companion animals is not the answer, though, for it just isn't fair to them. Instead, we humans must change our relationship with the non-human residents on our planet and learn to treat all living things with compassion and respect. I will continue to learn from all the animals that come into my life and to share this knowledge with others. Perhaps this is a small step in bringing us all back together.

Fortunately, science is now catching up with what those of us who have deeply connected with non-human animals. One such scientist, Gay Bradshaw, PhD, has created a new field called Trans-species psychology. The trans in trans-species signifies that there is no scientific basis for maintaining separate fields and models for animal and human psychology. This field of science is changing old thinking that non-human animals were thought to lack many attributes such as emotions, feelings, sophisticated cognitive abilities and other qualities that defined only by being human.

My wish is that this new awareness will change the way we treat animals. Give them respect and treat compassionately so they can live in ways that suit them instead of dominating and controlling their lives to suit us.